Black H●les

A Student Text

Third Edition

Black Holes

A Student Text
Third Edition

Derek Raine & Edwin Thomas

University of Leicester, UK

Imperial College Press

Published by

Imperial College Press
57 Shelton Street
Covent Garden
London WC2H 9HE

Distributed by

World Scientific Publishing Co. Pte. Ltd.
5 Toh Tuck Link, Singapore 596224
USA office: 27 Warren Street, Suite 401-402, Hackensack, NJ 07601
UK office: 57 Shelton Street, Covent Garden, London WC2H 9HE

Library of Congress Cataloging-in-Publication Data
Raine, Derek J., 1946–   author.
   Black holes : a student text / Derek Raine, University of Leicester, UK, Edwin Thomas,
University of Leicester, UK. -- 3rd edition.
      pages cm
   Includes bibliographical references and index.
   ISBN 978-1-78326-481-0 (hardcover : alk. paper) -- ISBN 978-1-78326-482-7 (pbk. : alk. paper)
   1. Black holes (Astronomy) I. Thomas, E. G. (Edwin George), 1937–   author. II. Title.
   QB843.B55R35 2014
   523.8'875--dc23

                                                                                    2014034022

British Library Cataloguing-in-Publication Data
A catalogue record for this book is available from the British Library.

Typeset by Stallion Press
Email: enquiries@stallionpress.com

Printed in Singapore

Preface to the Second Edition

New science can seem quite weird at first: Newton's mystical action-at-a-distance; Maxwell's immaterial oscillations in a vacuum; the dice-playing god of quantum mechanics. In due course however we come to accept how the world is and teach it to our students. Our acceptance, and theirs, comes principally through mastery of the hard details of calculations, not from generalised philosophical debate (although there is a place for that later).

Black holes certainly seem weird. We know they (almost certainly) exist, although no-one will ever 'see' one. And they appear to play an increasingly central role both in astrophysics and in our understanding of fundamental physics. It is now almost 90 years since the Schwarzschild solution was discovered, 80 years since the first investigations of the Schwarzschild horizon, 40 years on from the first singularity theorems and perhaps time that we can begin to dispel the weirdness and pass on something of what we understand of the details to our undergraduate students. This is what we attempt in this book. We have tried to focus on the aspects of black holes that we think are generally accessible to physics undergraduates who may not (or may) intend to study the subject further. Many of the calculations in this book can be done more simply using more sophisticated tools, but we wanted to avoid the investment of effort from those for whom these tools would be of no further use. Those who do go further will appreciate all the more the power of sophistication.

The presentation assumes a first acquaintance with general relativity, although we give a brief recapitulation of (some of) the main points in Chapter 1. The treatment is however very incomplete: we do not consider the Einstein field equations because we do not demonstrate here that the black hole geometries are solutions of the equations. In fact, very little prior knowledge of relativity is required to study the properties of given black hole spacetimes. Chapter 2 is devoted to classical spherically symmetric, or non-rotating (Schwarzschild) black holes in the vacuum and Chapter 3 to axially symmetric, or rotating (Kerr) ones. It is unfortunate that even simple calculations for Kerr black holes rapidly become algebraically complex. We have tried not to let this obscure the intriguing physics. We ask the reader to stick with it. After all, there may be a lot of it, but it is only elementary algebra. We give only a brief overview of charged black holes. These have played an important role as algebraically simpler models for many of the properties of rotating holes, and they are important as such in higher dimensions, but it is intrinsically difficult to maintain an interest in the physics of objects that probably do not exist, especially since we are going to treat the rotating holes in detail anyway.

In Chapter 4 we attempt to explain the quantum properties of black holes without recourse to quantum field theory proper. The calculations here are less rigorous than in the rest of the book (probably a gross understatement) but many variations on the standard theme are now available in textbooks, reviews and lecture notes on the internet and we see no merit in repeating these. We hope our approach is more useful than a crash course in quantum field theory. It does, of course, assume a more than superficial understanding of standard quantum theory. Chapter 6 closes the book with a brief review of black hole astrophysics in so far as it is relevant to the observation of black holes. For this second edition we have added Chapter 5 on wormhole metrics and time travel and a set of solutions to the problems. The new edition has also given us the opportunity to revise and clarify some of the text and problems and to add some new problems.

We are aware that we have omitted many contemporary topics in black hole physics, not least the properties of general black holes, perturbation of black holes and the role of black holes in string theories. We regard these as beyond the scope of the book (and in the last case of the expertise of the authors). We hope (and believe) that working out long-hand the details of what we do include will provide a firm foundation for those students who will go on to study such advanced topics and a firm understanding and appreciation of the properties of black holes for those who do not.

Preface to the Third Edition

This book is intended to present the main features of black holes at a level accessible to final year undergraduates. Gravity far away from a (non-rotating) black hole is no different from gravity at the same distance from a star of the same mass, but close in we discover the relativistic effects of strong gravitational fields that lead to new and exotic behaviour. These effects are important if we want to understand how black holes might be detected. Our approach was to accept that the various black hole metrics were solutions of Einstein's equations and to explore the motion of light and matter in these metrics. In any case, with modern algebraic computing packages, it is straightforward to verify that these black hole metrics are solutions of the vacuum field equations. In this edition we have revised the chapters of the second edition, some more so than others, and added new material and some new problems. Also, we have corrected some minor errors which were, regrettably, present in the second edition.

In the second edition we covered research into black holes up to the point where it stood in the late 1970s. Up to that time studies of black holes were confined to four dimensions (three space dimensions and one of time) with the cosmological constant set to zero. With these restrictions the number of different types of black hole solutions is limited: black holes are characterised by their mass, angular momentum and electrical charge. Once we allow extra spatial dimensions, an idea which has gained significance through string theory, and a non-zero cosmological constant, which is supported by cosmological observations, things become a great deal more complicated. The field equations that related the metric to the matter content are no longer unique and whichever ones we choose, they allow a plethora of black hole solutions.

In this new edition we have added a chapter in which we outline recent research on black holes in more than four spacetime dimensions and allows a non-zero (negative) cosmological constant. In keeping with the original philosophy of the book, we investigate the properties of various metrics, without demonstrating that they solve any field equations. However, whereas in the rest of the book we have been able to give detailed arguments, in this chapter we can do no more than outline some interesting examples. We hope however that the chapter justifies its inclusion in raising some of the most intriguing issues of fundamental physics. In particular, in the period after Hawking announced his discovery that black holes radiate, there were hopes that this might be the big clue to a quantum theory of gravity. It may still be the case, but, if so, unravelling the clue is taking a long time. We recognise that for many of our readers this will remain someone else's problem. For all of our students though we hope that this might provide a glimpse of what is at stake.

Derek Raine
Ted Thomas

Contents

Contents

Chapter 1

Relativistic Gravity

1.1 What is a black hole?

Black holes arise because gravity affects the way light waves travel through space. Newtonian dynamics does not treat the effect of gravity on light, but we can use our Newtonian intuition to guess what sort of interaction there might be, given that there is one. The propagation of light is governed by its speed c, so it is natural to look for a characteristic speed in Newtonian gravity. The obvious one is the escape velocity $v_{\mathrm{esc}} = (2GM/R)^{1/2}$ from a body of mass M and radius R. So we expect that light emitted from the body will fail to escape to a distant observer when $v_{\mathrm{esc}} > c$, that is when the mass of a body of density ρ exceeds the value

$$M \sim (c^2/G)^{3/2} \rho^{-1/2},$$

where we have used the approximation $\rho \sim M/R^3$.

Before the discovery in the twentieth century of white dwarf stars, and later neutron stars, the densest matter known was something like lead. So the mass M_* of the smallest object having a normal density, of around say $\rho_* \sim 5000\,\mathrm{kg\,m^{-3}}$, that could prevent its light from escaping to a distant observer would appear to be

$$M_* \sim 10^8 (\rho_*/\rho)^{1/2} M_\odot, \tag{1.1}$$

where $M_\odot$ is the solar mass. The mass M_* is very much greater than that of any known star. Such an object would be invisible or, as we would now say, would be a black hole.

These Newtonian arguments were known to Mitchell (the successor to Cavendish at Cambridge, and the first person to carry out Cavendish's experiment to measure G) and to Laplace, who was the first to carry out detailed studies of the many-body problem in the Solar System.

Two things have changed this picture. The first was the discovery of a relativistic theory of gravity, Einstein's general theory of relativity, in which the behaviour of light under the influence of gravity is treated unambiguously. The second was the discovery of matter having densities in the range 10^9 to 10^{17} kg m^{-3} in the form of white dwarfs and neutron stars. The existence of bodies of such high densities suggests that stellar mass black holes might exist. We can add to these another class of astronomical object, the active galactic nuclei (or AGNs), in which the central objects have normal densities but masses of the order of 10^8 $M_\odot$. This combination of mass and density make AGNs candidates for black holes.

In a relativistic theory of gravity, by definition, the local speed of light is always c, the speed of light in a vacuum in the absence of gravity. The Newtonian picture of the emitted light being slowed down and turned back by gravity is therefore not appropriate. However, we can get a truer picture from an application of the equivalence principle. The presence of a gravitational field introduces a relative acceleration between freely falling frames of reference. The equivalence principle then leads to an approximate relation between time interval $d\tau$ measured at radius R from a body of mass M and the corresponding time interval $d\tau'$ measured at infinity (Will, 1993)

$$d\tau \approx \left(1 - \frac{2GM}{Rc^2}\right)^{1/2} d\tau'. \qquad (1.2)$$

(The exact relation depends on the full theory of gravity.) This says that a clock at radius r runs slow compared to a clock at infinity. There is a corresponding redshift z of light emitted at frequency ω and received at ω' given by $\omega/\omega' = 1 + z = d\tau'/d\tau$. This suggests that $z \to \infty$ as the Newtonian potential $2GM/R \to c^2$ and hence that the light from the surface of a body at this potential would be redshifted to invisibility. Thus the body would be a black hole. The relativistic condition $2GM/R = c^2$ is, of course, analytically the same as the Newtonian condition $v_{esc} = c$. But the condition for the Newtonian approximation to be valid is that $GM/R \ll c^2$, so we require the full theory of gravity to investigate this behaviour consistently and to treat black holes correctly. Furthermore, the gravitational potential on the surface of a neutron star is about $0.1c^2$ so we need a general relativistic theory of stellar evolution to be confident of understanding evolution beyond this stage.

The replacement of Newton's theory of gravity by Einstein's general theory of relativity does not alter the relationship between mass and density in equation (1.1), except that now the density is to be interpreted as the average density within the boundary of the (non-rotating) black hole. But it does alter our picture of the spacetime of a black hole and how it gravitates.

A relativistic black hole has no material surface; all of its matter has collapsed into a singularity that is surrounded by a spherical boundary called its event horizon. The event horizon is a one-way surface: particles and light rays can enter the black hole from outside but nothing can escape from within the horizon of the hole into the external universe. An outgoing photon that originates outside the event horizon can propagate to infinity but in so doing it suffers a gravitational redshift: in Newtonian language it loses energy in doing work against the gravitational potential. This redshift is larger the closer the point of emission is to the horizon. On the other hand, a photon or particle emitted inside the horizon in any direction must inevitably encounter the singularity and be annihilated. (This is strictly true only in the simplest type of black hole: in more general black holes destruction is not inevitable and the fate of a particle or photon inside the hole is more complicated.) A photon emitted at the horizon towards a distant observer stays there indefinitely. For this reason one can think of the horizon as made up of outwardly directed photons.

1.2 Why study black holes?

The importance of black holes for gravitational physics is clear: their existence is a test of our understanding of strong gravitational fields, beyond the point of small corrections to Newtonian physics, and a test of our understanding of astrophysics, particularly of stellar evolution. The theory of stellar evolution, together with general relativity, shows that black holes are an almost inevitable consequence of the way that massive stars evolve. We therefore expect to find black holes amongst the stars in the Galaxy, and it appears that we do. The observational evidence for their existence is now very strong indeed. The gravitational waves emitted in the final stages of a black hole binary as it coalesces to form a single black hole would form a test of the predictions of general relativity.

There is also a surprising and quite unexpected reason why black holes turn out to be important: this is for the potential insight they offer into the connection between quantum physics and gravity. We shall see that black holes appear formally to satisfy the laws of thermodynamics, with Mc^2 in the role of internal energy, the acceleration due to gravity in the guise of temperature and the black hole area as entropy. But this turns out to be more than a formal analogy. When we include the effects of quantum physics we find that black holes behave as real objects with a non-zero temperature and entropy: in particular they radiate like black bodies. The analogy with thermodynamics is therefore not just a formal one, but black holes really do obey the laws of thermodynamics.

We can now turn this argument around. Since black hole radiation involves a mixture of gravity and quantum physics this connection necessarily leads us into the territory of quantum gravity, and, since quantum gravity is the missing link in a complete picture of the fundamental forces, to 'theories-of-everything'. Any theory-of-everything has to be consistent with thermodynamics, and hence with black hole thermodynamics. Therefore any theory-of-everything should be able to predict the thermodynamic properties of black holes from *ab initio* statistical calculations. It is therefore interesting that theories that treat 'strings' as fundamental entities have been partially successful in this regard. It appears that black holes will play a central role in our understanding of fundamental physics.

1.3 Elements of general relativity

It is assumed that the reader has had a first acquaintance with a course on general relativity, for example from one of the many excellent introductory textbooks (for example, Kenyon, 1990; Hartle, 2003). In this section we shall present some of the main ideas of the theory, but only in the form of a brief review.

1.3.1 *The principle of equivalence*

The principle of equivalence tells us that local experiments (those carried out in the immediate vicinity of an event) cannot distinguish between an accelerated frame of reference and the presence of a gravitational field. In both cases we observe that bodies subject to no non-gravitational forces fall with equal acceleration. (We use the double negative in 'no non-gravitational forces' to emphasise that gravity may or may not be present, but no other forces are acting.) This creates difficulties for the Newtonian approach to dynamics because that requires us to choose a non-accelerated (or 'inertial') frame of reference. In Newtonian physics we get round this problem by designating the distant stars as a non-accelerated reference frame. This is a non-local, non-causal solution and therefore unsatisfactory. It is obviously non-local, and it is non-causal because there is no mechanism by which this reference frame is singled out, except by the fact that it gives the right answers (for example, for the motion of the planets in the Solar System).

Einstein was struck by the observation that all bodies fall with the same acceleration in a gravitational field. Newtonian gravity offers no explanation for this *universality of free-fall*. So Einstein used the universality of free fall to enunciate the principle of equivalence and made this the basis of his general relativistic theory of gravity. The principle of equivalence implies that in free fall we cannot detect the presence of a gravitational field by local experiments. (In free fall all bodies move inertially whether or not gravity is present.) Therefore in a local freely falling frame of reference we already know the laws of physics in the presence of gravity: they are the same as if gravity were absent!

1.3.2 *The Newtonian affine connection*

From a practical point of view it is not very easy to use the principle of equivalence directly for calculations. This is because in the presence of gravity the local freely falling frame of reference is changing from event to event and we are not experienced at doing calculations in ever changing frames of reference. Rather we need to translate this point of view into a fixed, but arbitrary, reference frame. Although reference frames and coordinate systems are not the same thing (because the axes of a reference frame are not required to be tangents to coordinate lines), for the present purposes we shall ignore the distinction between them.

Imagine therefore that the system of coordinates $(\xi^0, \xi^1, \xi^2, \xi^3) = (\xi^\mu)$, $(\mu = 0, 1, 2, 3)$ corresponds momentarily to the natural choice of the freely-falling observer, with ξ^0 the Newtonian time (up to a factor of c). Suppose further that another set of coordinates (x^μ) are defined in a global patch (although not necessarily the whole) of spacetime. Each set of coordinates is given in terms of the other by $x^\mu = x^\mu(\xi^\nu)$ and $\xi^\mu = \xi^\mu(x^\nu)$, $(\mu, \nu = 0, 1, 2, 3)$. Note that on the left of these equations the variable stands for an independent coordinate and on the right for a function. The eliding of these distinct meanings by use of the same symbol is common practice and useful for keeping track of dependencies provided that care is taken.

According to Newtonian physics a test body subject to no non-gravitational forces will move along a spacetime trajectory defined by

$$\frac{d^2\xi^\mu}{d\tau^2} = 0. \qquad (1.3)$$

Transforming to our (x^μ) coordinates this becomes, after some calculation of partial derivatives,

$$\frac{d^2 x^\mu}{d\tau^2} + \frac{\partial x^\mu}{\partial \xi^\alpha} \frac{\partial^2 \xi^\alpha}{\partial x^\nu \partial x^\rho} \frac{dx^\nu}{d\tau} \frac{dx^\rho}{d\tau} = 0, \tag{1.4}$$

where we are employing the usual summation convention implying a sum over repeated indices. The Greek indices are assumed to range over the values 0,1,2,3 throughout. To derive (1.4) from equation (1.3) we have used

$$\frac{\partial x^\alpha}{\partial \xi^\mu} \frac{\partial \xi^\mu}{\partial x^\beta} = \delta^\alpha_\beta, \tag{1.5}$$

which follows from differentiation of $x^\alpha(\xi^\mu(x^\beta)) = x^\alpha$. equation (1.4) is of the form

$$\frac{d^2 x^\mu}{d\tau^2} + \Gamma^\mu_{\nu\rho} \frac{dx^\nu}{d\tau} \frac{dx^\rho}{d\tau} = 0. \tag{1.6}$$

This equation for the motion of a test body holds whether or not gravity is present. If gravity is present the only place it can appear in this equation is through the quantity $\Gamma^\mu_{\nu\rho}$, called the affine connection. Looked at in this way, gravity therefore does not enter through an additional force term on the right-hand side of (1.3). The only difference between the presence or absence of a gravitational field is that in the latter case it will be possible to recover the equation (1.3) everywhere in a single inertial coordinate system, not just locally in a freely-falling frame (because in the absence of gravity a local freely-falling frame is automatically a global inertial frame).

1.3.3 *Newtonian gravity*

To make the connection with the usual form of the equation of motion of a test body in Newtonian gravity we must be able to choose coodinates in which the affine connection takes an appropriate form. In fact, we must have, in some suitably chosen frame of reference (x^μ),

$$\Gamma^0_{ij} = 0; \quad \Gamma^i_{0j} = 0; \quad \Gamma^i_{00} = \frac{\partial \phi}{\partial x^i},$$

where $i, j = 1, 2, 3$, and ϕ is the Newtonian gravitational potential, since with these values for $\Gamma^\mu_{\nu\rho}$ we recover from (1.5) the equations of motion in a Newtonian gravitational field:

$$x^0 = ct = c\tau;$$

$$\frac{d^2 x^i}{d\tau^2} = -\frac{\partial \phi}{\partial x^i}.$$

This also tells us that the affine connection is related to the distribution of matter through the extension to a general coordinate system of Poisson's equation $\nabla^2 \phi = 4\pi G \rho$. Since this involves second derivatives of ϕ the relation between the affine connection and matter must involve the derivatives of the affine connection. The appropriate combinations of derivatives can be shown to be related to the curvature of (Newtonian) spacetime.

Newtonian gravitation is therefore a theory of the structure of spacetime, the relevant structures being the affine connection, the privileged time coordinate t and the Euclidean spatial metric. In Newtonian physics the (affine) geometry of spacetime is measured by the paths of particles and is unrelated to the geometry of time and space as measured by clocks and rods. Relativity is a lot simpler: there is only one geometry. The geometry of time and space, as measured by clocks and rods, itself governs the motion of particles, as we shall explain below.

1.3.4 *Metrics in relativity*

In a freely-falling frame special relativity is valid *locally* (whether or not gravity is present), and the spacetime interval (the proper distance or proper time) between neighbouring events, ds, is given by the familiar line element (or metric)

$$ds^2 = c^2 dt^2 - (dx^2 + dy^2 + dz^2), \tag{1.7}$$

where t, x, y, z are the time and rectangular spatial coordinates of the freely falling (inertial) observer. In tensor notation this line element can be written

$$ds^2 = \eta_{\alpha\beta} d\xi^\alpha d\xi^\beta, \tag{1.8}$$

where $\eta_{\alpha\beta}$ is the metric tensor and we are using $(\xi^\alpha) = (ct, x, y, z)$ for this special coordinate system at a point. With these coordinates the metric has components $\eta_{\alpha\beta} = \mathrm{diag}(1, -1, -1, -1)$. In this convention ds^2 is positive for a timelike interval, zero for a lightlike interval and negative for a spacelike interval. This is convenient for dealing with the motion of particles. For positive ds^2 then $ds/c = d\tau$, where $d\tau$ is the proper time between the events. For negative ds^2 then $(-ds^2)^{1/2} = dl$ is the proper distance between the two events. An alternative convention often used for the metric is $\eta_{\alpha\beta} = \text{diag}(-1,1,1,1)$. Care is needed in copying formulae from references to adjust these, if necessary, to the convention being employed.

The coordinate labels (t,x,y and z) we attach to events are arbitrary (subject to some reasonable continuity and distinguishability conditions), but the proper time (measured by clocks) and proper distance (measured by rulers) are physical quantities and so must be independent of the coordinate labels. We should therefore see what happens to the metric if we transform to new coordinates. In a general coordinate system, which we shall call (x^μ), we have

$$ds^2 = \eta_{\alpha\beta}\frac{\partial \xi^\alpha}{\partial x^\mu}\frac{\partial \xi^\beta}{\partial x^\nu}dx^\mu dx^\nu = g_{\mu\nu}dx^\mu dx^\nu \qquad (1.9)$$

(from the chain rule for partial derivatives, $d\xi^\alpha = (\partial \xi^\alpha/\partial x^\mu)dx^\mu$). The quantities $g_{\mu\nu}$ are called the metric coefficients. They are symmetric, that is to say $g_{\mu\nu} = g_{\nu\mu}$, and are, in general, functions of the spacetime coordinates.

In the absence of gravity we can find a *global* coordinate system (ξ^α) in which the metric takes the form (1.8) everywhere. In the presence of gravity we can find such coordinates only in an infinitesimal neighbourhood of each spacetime point.

Problem 1.1. *Show that neither the Minkowski metric equation (1.7) nor the sum of squares is an invariant under the Galilean transformation and hence that there is no interval in Newtonian spacetime (i.e. Newtonian spacetime does not admit a metric).*

The motion of particles is again governed by equation (1.6) in relativity, because the arguments leading to it are still valid, except that now both $t \neq \tau$ and $x^0 \neq c\tau$ (since these would not be compatible with the metric equation (1.8) or (1.9)).

Now we see that in local freely falling coordinates the metric is obtained from first derivatives of the ξ^μ whereas the affine connection depends on second derivatives. We therefore expect that there will exist a relation between the components of the affine connection and derivatives of components of the metric. This is indeed the case:

$$\Gamma^\mu_{\nu\rho} = \frac{1}{2}g^{\mu\alpha}(\partial_\nu g_{\rho\alpha} + \partial_\rho g_{\nu\alpha} - \partial_\alpha g_{\nu\rho}), \qquad (1.10)$$

where $\partial_\alpha f = \partial f / \partial x^\alpha$ and $(g^{\mu\nu})$ is the inverse matrix to $(g_{\mu\nu})$, so

$$g^{\mu\alpha} g_{\alpha\nu} = \delta^\mu_\nu.$$

Thus we find the values of $\Gamma^\mu_{\nu\rho}$ from the values of the metric coefficients appropriate to a particular gravitational field.

So in relativity there is just one spacetime geometry defined by the metric and measured by both clocks and rods *and* by particle paths. The motion of particles is governed by equation (1.6), called the geodesic equation (because it is of the same form as the equation for the shortest paths on a curved surface, which are called geodesics). Note that the physical interpretation of the coordinates (x^μ) (what they measure physically) depends on the metric coefficients, so cannot be determined unless the metric is known.

1.3.5 *The velocity and momentum 4-vector*

In this section we define two important 4-vectors relating to the motion of a particle, namely its velocity and momentum; acceleration will be dealt with later. The 4-velocity vector of a particle with position $(x^\alpha(\tau))$ is given by

$$(u^\mu) = \left(\frac{dx^\mu}{d\tau} \right) = \left(\frac{dx^0}{d\tau}, \frac{dx^1}{d\tau}, \frac{dx^2}{d\tau}, \frac{dx^3}{d\tau} \right),$$

formally as in special relativity. The 4-momentum of a particle of mass m_0 is $p^\mu = m_0 u^\mu$.

We lower indices using the metric tensor $g_{\mu\nu}$, so the covariant components of momentum are $p_\mu = g_{\mu\nu} p^\nu$. Conversely, we raise indices using $g^{\mu\nu}$, so the contravariant components are $p^\mu = g^{\mu\nu} p_\nu$.

Taking the scalar product of the 4-velocity with itself gives a relation that we shall use repeatedly below:

$$u^\mu u_\mu = c^2, \tag{1.11}$$

and, similarly,

$$p^\mu p_\mu = m_0^2 c^2. \tag{1.12}$$

Recall that a scalar produce is independent of the frame of reference in which it is evaluated. In particular, we can choose the local freely falling frame of the particle. In this frame special relativity is valid, so the metric takes the form (1.7) and the components of (u^μ) are $(c, 0, 0, 0)$. Thus in this frame we readily verify equation (1.11) and similarly equation (1.12).

1.3.6 *General vectors and tensors*

By definition, if we make a coordinate transformation $x \to x'(x)$, then the contravariant components of a 4-vector v^μ transform according to

$$v'^\mu = \frac{\partial x'^\mu}{\partial x^\nu} v^\nu$$

and the covariant components as

$$v'_\mu = \frac{\partial x^\nu}{\partial x'^\mu} v_\nu,$$

with

$$\frac{\partial x'^\mu}{\partial x^\nu} \frac{\partial x^\nu}{\partial x'^\rho} = \delta^\mu_\rho.$$

Higher rank tensors (having multiple indices) transform by extension of this rule to each index. For example:

$$g'_{\mu\nu} = \frac{\partial x^\alpha}{\partial x'^\mu} \frac{\partial x^\beta}{\partial x'^\nu} g_{\alpha\beta}.$$

Problem 1.2. *Obtain the inverse transformations*

$$v^\mu = \frac{\partial x^\mu}{\partial x'^\nu} v'^\nu \quad and \quad v_\mu = \frac{\partial x'^\nu}{\partial x^\mu} v'_\nu.$$

Problem 1.3. *By showing that $v_\mu v^\mu = v'_\mu v'^\mu$ verify the invariance of the scalar product.*

1.3.7 *Locally measured physical quantities*

We often need to calculate the energy or velocity of a particle as measured by a local observer, for example, an observer having a fixed location in some global coordinate system, or a locally freely falling observer. The energy of a particle with respect to a local observer with 4-velocity u^μ_{obs} is the time component of the 4-momentum of the particle in the observer's frame of reference and is therefore obtained by projecting the 4-momentum on to the velocity 4-vector of the observer. So

$$\mathcal{E} = u^\mu_{\text{obs}} p_\mu = m_0 \gamma c^2,$$

where $\gamma = (1 - v^2/c^2)^{-1/2}$ is the usual relativistic gamma factor and v the local velocity of the particle. As in Section 1.3.5, we have evaluated the scalar product in the local freely falling frame of the observer where special relativity applies. The scalar product is an invariant, that is, its value is independent of the coordinate system used to evaluate it, so, for example, we can evaluate it in a local inertial frame and also in a global coordinate system and thus relate the global energy to the locally measured energy. We will make frequent use of this property of the scalar product in subsequent chapters.

Physical quantities in two local frames at the same event will be connected by a Lorentz transformation between them even though one or both of the frames may be accelerating. This follows because the instantaneous rates of clocks and lengths of rods are not affected by accelerations and depend only on the relative velocities.

1.3.8 *Derivatives in relativity*

Given a vector field $v^\mu(x)$, say, it is easy to see that the usual partial derivative $\partial v^\mu/\partial x^\nu$ is, in general, not a physically meaningful quantity: we can give it any value we like by the choice of coordinates x^μ. In particular, a vector field which is constant in one coordinate system, in which its partial derivatives will vanish, may have non-zero derivatives in another coordinate system. The problem arises because the usual definition of a partial derivative compares the values of the quantity in question at two different points and hence, in general, at points where the coordinates can be changed independently. Rates of change having a physical interpretation will be formed from differences in quantities at the same point. Various such derivatives exist, depending on how the quantities are brought to the same point.

The most common, although not the simplest, is the (so called) covariant derivative. This is a generalisation of the behaviour of a tangent vector to a curve. Let $V^\mu = dx^\mu/d\tau$ be the tangent vector to the curve $x^\mu = x^\mu(\tau)$. Equation (1.6) tells us that

$$\delta V^\mu = -\Gamma^\mu_{\nu\rho} V^\nu \delta x^\rho$$

is the change in the tangent vector on going from x^μ to $x'^\mu = x^\mu + \delta x^\mu$ along a geodesic. We now use this to define the parallel transport of any vector, A^μ say, between two neighbouring points. To form the covariant derivative we take the difference between the vector A^μ at x' and the parallelly transported

$A^\mu + \delta A^\mu = A^\mu - \Gamma^\mu_{\nu\rho} A^\nu \delta x^\rho$ at x', in the usual limit:

$$\nabla_\nu A^\mu = \lim_{\delta x \to 0} \frac{A^\mu(x + \delta x) - (A^\mu(x) + \delta A^\mu)}{\delta x^\nu}$$

$$= \frac{\partial A^\mu}{\partial x^\nu} + \Gamma^\mu_{\nu\rho} A^\rho.$$

The quantity $\nabla_\nu A^\mu$ is the covariant derivative of A^μ, also denoted by a semi-colon as $A^\mu_{;\nu}$, or sometimes by DA^μ/dx^ν. The covariant derivative therefore measures the derivative corresponding to parallel transport. Note that in this notation the condition for a geodesic becomes that the tangent vector should have zero covariant derivative along the tangent, so $V^\nu \nabla_\nu V^\mu = 0$. This is the total or directional derivative, also written in terms of the affine parameter τ as $DV^\mu/d\tau = 0$.

Problem 1.4. *The covariant derivative of a scalar is just the usual (partial) derivative (because the value of a scalar is independent of the coordinate system). Hence, for two vector fields, A^μ and B_μ, we have $\nabla_\nu(A^\mu B_\mu) = \partial_\nu(A^\mu B_\mu)$. Use this to show that*

$$\nabla_\nu B_\mu = \frac{\partial B_\mu}{\partial x^\nu} - \Gamma^\rho_{\mu\nu} B_\rho.$$

Since we have seen that in a freely falling frame of reference the components of the affine connection vanish at a point, another definition of the covariant derivative is that it reduces to the ordinary (partial) derivative in a freely falling frame. This suggests an algorithm for implementing the equivalence principle: in a local freely falling Cartesian frame the laws of physics take the same form as in the absence of gravity and so involve ordinary derivatives in these frames. To introduce the effect of gravity in a general coordinate frame we therefore replace ordinary derivatives by covariant derivatives and the Minkowski metric by $g_{\mu\nu}$.

For example, in Chapter 4 we shall need the scalar wave equation in a gravitational field. In the absence of gravity we have, for a field ψ, the usual wave equation

$$\frac{\partial^2 \psi}{\partial t}^2 = \nabla^2 \psi$$

where ∇^2 is the usual Laplacian operator in three dimensions. The wave equation can be written more compactly as

$$\eta^{\mu\nu} \frac{\partial}{\partial x^\mu} \frac{\partial}{\partial x^\nu} \psi = 0.$$

In the presence of gravity this becomes

$$g^{\mu\nu}\nabla_\mu\nabla_\nu\psi = 0.$$

However, this prescription works only if gravity does not introduce an explicit curvature term (derivatives of the connection) into the equation.

A second type of derivative that is sometimes useful is the Lie derivative. To define this we use for the transported vector field at $x'^\mu = x^\mu + V^\mu\delta\tau$ the vector obtained by using the standard transformation law between coordinate systems, namely,

$$A^\mu + \delta A^\mu = \frac{\partial x'^\mu}{\partial x^\nu} A^\nu$$

$$= \left(\delta^\mu_\nu + \frac{\partial V^\mu}{\partial x^\nu}\delta\tau\right) A^\nu.$$

Note that the transformation is here regarded as an active one, taking the point x^μ to a different point x'^μ, rather than a passive change of coordinates at a given point. The Lie derivative is then defined as

$$\mathcal{L}_V A^\mu = \lim_{\delta\tau\to 0} \frac{A^\mu(x + \delta x) - (A^\mu(x) + \delta A^\mu)}{\delta\tau}$$

$$= V^\nu\partial_\nu A^\mu - A^\nu\partial_\nu V^\mu.$$

Problem 1.5. *Show that the partial derivatives in the definition of the Lie derivative can be replaced by covariant derivatives, i.e. that*

$$\mathcal{L}_V A^\mu = V^\nu\nabla_\nu A^\mu - A^\nu\nabla_\nu V^\mu.$$

Roughly speaking we can say that the Lie derivative is the directional derivative of a vector field along a curve adjusted for the change in the tangent.

Problem 1.6. *Generalise the tensor transformation law, the covariant derivative and the Lie derivative to a tensor field $T^{\mu\nu}$.*

We shall give an example of the use of the Lie derivative below (problem 1.8).

1.3.9 *Acceleration 4-vector*

As an example of the covariant derivative we can construct the acceleration 4-vector in a general coordinate system. In special relativity, in Minkowski coordinates, the acceleration is obtained from the velocity

4-vector $(u^\mu) = (dx^\mu/d\tau)$ as $(a^\mu) = (du^\mu/d\tau)$. In the presence of gravity we replace the ordinary derivative by a covariant derivative as follows:

$$a^\mu = \frac{Du^\mu}{d\tau} = \frac{dx^\alpha}{d\tau}\frac{Du^\mu}{dx^\alpha}$$

$$= \frac{dx^\alpha}{d\tau}\left(\frac{\partial u^\mu}{\partial x^\alpha} + \Gamma^\mu_{\beta\alpha}u^\beta\right)$$

$$= \frac{du^\mu}{d\tau} + \Gamma^\mu_{\beta\alpha}u^\beta u^\alpha.$$

The magnitude of the rest frame or proper acceleration a of a particle is given by

$$a^\mu a_\mu = -a^2. \tag{1.13}$$

1.3.10 *Paths of light*

In relativity the path of a light ray is also governed by the metric through

$$0 = g_{\mu\nu}dx^\mu dx^\nu$$

and

$$\frac{d^2x^\mu}{d\lambda^2} + \Gamma^\mu_{\beta\gamma}\frac{dx^\beta}{d\lambda}\frac{dx^\gamma}{d\lambda} = 0,$$

in which λ is a parameter which varies along the world line of a light ray (and is called an affine parameter, because it maintains for light rays the usual form of the geodesic equation).

1.3.11 *Einstein's field equations*

The problem of relating the metric coefficients $g_{\mu\nu}(x^\alpha)$, as functions of the coordinates, to the distribution of matter is solved by Einstein's field equations, which are

$$R_{\mu\nu} - \frac{1}{2}g_{\mu\nu}R + \Lambda g_{\mu\nu} = \frac{8\pi G}{c^4}T_{\mu\nu}. \tag{1.14}$$

Here $R_{\mu\nu}$ is the Ricci curvature tensor given by

$$R_{\mu\nu} = \frac{\partial\Gamma^\gamma_{\mu\nu}}{\partial x^\gamma} - \frac{\partial\Gamma^\gamma_{\mu\gamma}}{\partial x^\nu} + \Gamma^\gamma_{\mu\nu}\Gamma^\delta_{\gamma\delta} - \Gamma^\delta_{\nu\gamma}\Gamma^\gamma_{\mu\delta} \tag{1.15}$$

and $R = g^{\mu\nu} R_{\mu\nu}$ is the Ricci scalar. The tensor $T_{\mu\nu}$ is the energy momentum tensor which includes all the sources of the gravitational field excluding the energy and momentum in the gravitational field itself which is accounted for by the non-linearity of the equations. The constant Λ is called the cosmological constant. Equations (1.14) provide ten non-linear partial differential equations for the metric coefficients. In Chapters 2 and 3 we shall be studying black hole solutions of the Einstein field equations in a vacuum, for which $T_{\mu\nu} = 0$. We shall also put $\Lambda = 0$. Observations from cosmology appear to give Λ a small positive value. In terms of a length scale, the value is of order $\Lambda^{-\frac{1}{2}} = 10^{26}$m. Thus, on non-cosmological scales we are justified in ignoring the cosmological constant. In Section 4.7 we shall need the full equations (1.14). There are also some key theoretical results that take Λ to have a negative value which we shall discuss in Section 5.8.

There is one further remark we can usefully make that will help in understanding some of the more recent developments in later chapters. This is that the field equations can be derived from an action principle: specifically, any metric that extremises the integral

$$S = \frac{1}{16\pi G} \int \sqrt{-g}(R + 2\Lambda)d^4x \qquad (1.16)$$

is a solution of Einstein's equations (with a cosmological constant) in vacuo. The quantity S in (1.16) is called the Einstein (or Hilbert–Einstein) action. In four spacetime dimensions the Einstein action gives the most general form of the field equations that contain no derivative of the metric higher than second order and hence reduce to Newtonian gravity for gravitational weak fields.

Problem 1.7. *Show that if $T_{\mu\nu} = 0$ then $R = 0$.*

1.3.12 *Solving Einstein's field equations*

Einstein's equations form a set of coupled non-linear partial differential equations for which there are no systematic methods of solution for given matter distributions. (If there are no constraints on the matter, then the energy momentum tensor can be anything and any metric is a solution for some $T_{\mu\nu}$.) Nevertheless a large number of solutions have been found. These are usually obtained by restricting the form of the metric coefficients $g_{\mu\nu}$ by considerations of symmetry, thereby simplifying the field equations. Nevertheless, computing by hand the components of the Einstein equations from even a quite simple metric form is a major undertaking. Fortunately, mathematical software is now available that greatly simplifies the task. For example, in MAPLE the *tensorsGR* package within *tensor* contains a set of commands to evaluate the curvature tensors (and hence the field equations) for any given functional form of the metric. One can use this readily to check that the metrics in this book are indeed solutions.

1.3.13 *Symmetry and Killing's equation*

Despite the complexity of Einstein's equations a surprising number of exact solutions are known. These are obtained by imposing symmetries on the spacetime which restrict the possible form of the metric. We shall meet two cases in the following, namely spherical and axial symmetry. In order to exploit these symmetries, we shall need one mathematical result which we derive here.

Let the vector k^μ be a direction of symmetry; for example k^μ might point along the azimuthal (ϕ -) direction in spherical symmetry, i.e. $(k^\mu) = (0,0,0,1)$ in spherical polar coordinates. Then we shall show that

$$k_{\mu;\nu} + k_{\nu;\mu} = 0, \tag{1.17}$$

where the semi-colon denotes the covariant derivative

$$k_{\mu;\nu} = \frac{\partial k_\mu}{\partial x^\nu} - \Gamma^\lambda_{\mu\nu} k_\lambda.$$

Equation (1.17) is called Killing's equation.

Let events P at (x^μ) and Q at (x'^μ) be separated by a small distance in the direction of symmetry k^μ, so

$$x'^\mu = x^\mu + \varepsilon k^\mu. \tag{1.18}$$

In moving from P to Q we do two things: we change the label of the point from x^μ to $x^\mu + \varepsilon k^\mu$ and we make an active transformation of coordinates from x^μ to x'^μ. The metric coefficients at x'^μ are related to those at x^μ by

$$g_{\mu\nu}(x^\lambda) = \frac{\partial x'^\alpha}{\partial x^\mu} \frac{\partial x'^\beta}{\partial x^\nu} g'_{\alpha\beta}(x'^\lambda).$$

As we have moved in a direction of symmetry the *function* $g'_{\mu\nu}(x'^\lambda)$ of the new coordinates has the same form as the *function* $g_{\mu\nu}(x^\lambda)$ of the original coordinates, so we can substitute

$$g'_{\mu\nu}(x'^\lambda) = g_{\mu\nu}(x'^\lambda).$$

Expanding $g_{\alpha\beta}(x'^\lambda)$ as a Taylor series and using equation (1.18) to evaluate the partial derivatives we get

$$g_{\mu\nu}(x^\lambda) = (\delta_\mu^\alpha + \varepsilon\partial_\mu k^\alpha)(\delta_\nu^\beta + \varepsilon\partial_\nu k^\beta)(g_{\alpha\beta}(x^\lambda) + \varepsilon k^\lambda\partial_\lambda g_{\alpha\beta}(x^\lambda) + \cdots).$$

Multiplying out the brackets to first order in ε gives us

$$g_{\mu\nu}(x^\lambda) = g_{\mu\nu}(x^\lambda) + \varepsilon k^\lambda\partial_\lambda g_{\mu\nu}(x^\lambda) + \varepsilon g_{\mu\beta}(x^\lambda)\partial_\nu k^\beta + \varepsilon g_{\alpha\nu}(x^\lambda)\partial_\mu k^\alpha.$$

But

$$\partial_\nu(g_{\mu\beta}k^\beta) = \partial_\nu k_\mu = g_{\mu\beta}\partial_\nu k^\beta + k^\beta\partial_\nu g_{\mu\beta},$$

and we get

$$0 = \varepsilon(\partial_\mu k_\nu - k^\alpha\partial_\mu g_{\alpha\nu}) + \varepsilon(\partial_\nu k_\mu - k^\beta\partial_\nu g_{\mu\beta}) + \varepsilon k^\lambda\partial_\lambda g_{\mu\nu}$$
$$= \varepsilon(k_{\mu;\nu} + k_{\nu;\mu}),$$

where the final equality is obtained using the relation equation (1.10). A vector field k^μ satisfying this relation is called a Killing vector.

Problem 1.8. *Show that Killing's equation* (1.17) *is equivalent to* $\pounds_k g_{\mu\nu} = 0$. *(You will need the result of problem 1.6.)*

Chapter 2

Spherical Black Holes

In this chapter we look at the vacuum gravitational field produced by a spherical (non-rotating) mass. The spacetime metric for this case, obtained by Karl Schwarzschild in 1916, was the first exact solution of Einstein's equations to be found, although its properties were not fully understood until much later.

We investigate the geometry in the vicinity of the mass M by studying the motion of test particles and light rays. If the space is a vacuum down to the Schwarzschild radius $r = 2GM/c^2$ we obtain the metric of a spherical black hole. In the next chapter we shall study the metric of a rotating (non-spherical) black hole. The surface $r = 2GM/c^2$ constitutes the event horizon of the (spherical) black hole and acts as a one-way membrane for particles and light. We explore the motion of particles and light rays as a spherical mass collapses to form a black hole and what an observer falling into a spherical black hole would experience. To explore the geometry in the vicinity of the event horizon and the maximum extension of the geometry from the external region we introduce various alternative coordinate systems.

2.1 The Schwarzschild metric

The general spherically symmetric metric can be written in the form

$$ds^2 = g_{00}(t,r)c^2dt^2 + g_{11}(t,r)dr^2 - r^2(d\theta^2 + \sin^2\theta d\phi^2). \qquad (2.1)$$

We adopt the timelike convention for the signs of the metric coefficients $(+, -, -, -)$, so $ds^2 > 0$ is a timelike interval and $ds^2 < 0$ is a spacelike interval (Section 1.3.4). The metric contains two unknown functions of r and t, g_{00} and g_{11}, to be determined from Einstein's field equations.

In the vacuum of space surrounding a massive spherical body Einstein's equations force the functions to be independent of time, and to take the form

$$g_{00} = \left(1 - \frac{2GM}{c^2 r}\right), \tag{2.2}$$

$$g_{11} = -\left(1 - \frac{2GM}{c^2 r}\right)^{-1}, \tag{2.3}$$

where M is a parameter and G the gravitational constant. Thus the metric becomes

$$ds^2 = \left(1 - \frac{2GM}{rc^2}\right) c^2 dt^2 - \left(1 - \frac{2GM}{rc^2}\right)^{-1} dr^2 - r^2(d\theta^2 + \sin^2\theta d\phi^2). \tag{2.4}$$

This is the Schwarzschild metric in Schwarzschild coordinates (t, r, θ, ϕ). It describes the spacetime in the exterior vacuum region only. Inside a material body the metric is different and depends on the properties of the body such as the equation of state of the material.

The contravariant components of the metric $g^{\mu\nu}$ satisfy

$$g^{\mu\nu} g_{\nu\rho} = \delta^\mu_\rho,$$

defining $(g^{\mu\nu})$ as the matrix inverse to $(g_{\nu\rho})$. Since the metric is diagonal the matrix inverse is readily found, giving

$$g^{00} = \left(1 - \frac{2GM}{c^2 r}\right)^{-1},$$

$$g^{11} = -\left(1 - \frac{2GM}{c^2 r}\right),$$

$$g^{22} = -\frac{1}{r^2},$$

$$g^{33} = -\frac{1}{r^2 \sin^2\theta}.$$

In relativity the interpretation of the coordinates is obtained from the properties of the metric. We give the interpretation of the coordinates and of the parameter M in the next section.

2.1.1 *Coordinates*

The notation has been chosen here to be suggestive. The coordinate r does indeed play the role of a radial measure; in fact, the sphere $r = $ constant,

$t =$ constant has the (spacelike) metric

$$dl^2 = r^2(d\theta^2 + \sin^2 \theta d\phi^2), \tag{2.5}$$

from which it is clear that the element of area on the surface is $rd\theta \times r\sin\theta d\phi$, and hence that the area of the surface is $4\pi r^2$. Thus the r coordinate is related to the proper area, that is the area as measured physically by, say, the diminution in intensity of a wavefront of light. The angles θ and ϕ are the usual spherical polar angles. Nevertheless, in relativity one should always be aware that a suggestive notation can sometimes be misleading. Later on we shall see that there are regions of spacetime (those within $r = 2GM/c^2$) where r is not a spacelike coordinate and t is not timelike. We shall also find that it is sometimes convenient to use other coordinates to describe the Schwarzschild spacetime geometry.

2.1.2 *Proper distance*

The radial increment of proper distance is obtained from the metric by setting $dt = d\theta = d\phi = 0$; it is not dr, but $dr/(1 - 2GM/rc^2)^{1/2}$. The coordinate r therefore does *not* measure proper distance, as would be determined, for example, by a ruler.

2.1.3 *Proper time*

Similarly, the time coordinate t is *not* the time kept by a standard clock in the exterior gravitational field of the body. We can see this from the metric (2.4). An increment of proper time kept by a stationary clock at a radius r, for which $dr = d\theta = d\phi = 0$, is related to the increment of coordinate time by

$$d\tau = \left(1 - \frac{2GM}{rc^2}\right)^{1/2} dt. \tag{2.6}$$

The increments $d\tau$ and dt increasingly agree as $r \to +\infty$. Thus the time coordinate t has the interpretation that it is the proper time kept by a clock at an infinite distance from the body, or a close approximation to that time kept by clocks at large distances.

For the purpose of visualisation, we can imagine that the body is surrounded by static shells at each value of the radial coordinate r. A standard clock fixed to a shell measures proper time on that shell, and, of course, standard clocks are identical in that they run at the same rate when they are at the same radius. In addition, we imagine a second clock on each shell that has been adjusted to run at the same rate as the distant clock. This second group of clocks measure coordinate time and run faster than the standard clocks by a factor $(1 - 2GM/rc^2)^{-1/2}$. This slowing down of local clocks relative to a distant clock is the effect of gravitational time dilation. The closer we approach $r = 2GM/c^2$, the faster the coordinate clock runs relative to local proper time, until at $r = 2GM/c^2$ it is running infinitely faster.

The Schwarzschild coordinates (t, r, θ, ϕ) provide a global reference frame that enables us to evaluate quantities in terms of the distance and time measurements of an observer at infinity. As we shall see, the values obtained by a distant observer for time intervals, for energies and for velocities of particles differ from local measurements of these quantities.

2.1.4 *Redshift*

A consequence of this difference in the measurement of time locally and at infinity is that radiation sent out from a radius r is redshifted when received by a static observer further out. Since the wavelength of radiation is proportional to the period of vibration, equation (2.6) gives us

$$\lambda = \left(1 - \frac{2GM}{rc^2}\right)^{1/2} \lambda_\infty,$$

for the relation between the wavelength λ of radiation emitted at r and the wavelength λ_∞ received at infinity. For the redshift z, defined by

$$z = \frac{\lambda_\infty - \lambda}{\lambda},$$

we get

$$1 + z = \left(1 - \frac{2GM}{rc^2}\right)^{-1/2}.$$

2.1.5 *Interpretation of M and geometric units*

The quantity M in the Schwarzschild metric arises as a constant of integration in the solution of the Einstein equations. It is in fact the mass of the central gravitating body. This is determined by a comparison of the motion of a body in the Schwarzschild metric with the motion of a body in the Newtonian gravitational field of a mass M, which we shall carry out below. For convenience we define a *geometric mass* m,

$$m = \frac{GM}{c^2}. \tag{2.7}$$

The geometric mass has the dimension of a length, so in SI units, M would be given in kilograms and m in metres.

To simplify further, we replace ct by t and $c\tau$ by τ. The new time is therefore also measured in length units, hence, if we use SI units, in metres. Thus, we have t(in metres) $= c \times t$(in seconds), and similarly for τ, but instead of this cumbersome notation we allow the context to distinguish which t and τ we mean. In geometrical units the metric is

$$d\tau^2 = \left(1 - \frac{2m}{r}\right) dt^2 - \left(1 - \frac{2m}{r}\right)^{-1} dr^2 - r^2(d\theta^2 + \sin^2\theta d\phi^2). \tag{2.8}$$

2.1.6 *The Schwarzschild radius*

The radial coordinate value $r = 2GM/c^2 = 2m = r_S$ is called the Schwarzschild radius. We see that something strange appears to happen to the Schwarzschild metric at this value of r, since the metric coefficient g_{11} becomes infinite. For a normal gravitating body the Schwarzschild radius lies within the object where the vacuum Schwarzschild solution ceases to apply. For the Sun, for example, we get $r_S \sim 3$ km. So, for the dynamics of the Solar System, or even for motion around a neutron star, the problem can be safely ignored. But the question of what happens at the Schwarzschild radius leads to the subject of black holes. Thus, if a spherical mass has radius $R > r_S$, then the Schwarzschild metric in the region $r > R$ gives its exterior gravitational field; if the region $r \geq r_S$ is a vacuum then the Schwarzschild metric for $r > r_S$ gives the exterior gravitational field of a black hole.

Inside $r = r_S$ the Schwarzschild metric is again a solution of Einstein's vacuum field equations and gives the interior geometry of a black hole. As we shall show later, the reason that we cannot follow what happens at the Schwarzschild radius $r = r_S$ in Schwarzschild coordinates is because the coordinates are invalid at this surface, not because of any physical limitation.

At $r = 0$ the coefficient of g_{00} becomes infinite. In contrast to the apparent singularity at the Schwarzschild radius, this singularity cannot be removed by a different choice of coordinates. It is a physical singularity where spacetime curvature becomes infinite and timelike worldlines terminate.

2.1.7 *The event horizon*

As we shall see,.of the many interesting properties of the surface $r = r_S$ the most novel is that it acts as a one-way membrane for information: things can fall into the region $r < r_S$ but cannot emerge from it. For this reason it is called an event horizon and the region of space within it is called a black hole.

Problem 2.1. *Show that the (proper) circumference of the event horizon of a black hole of mass m is $4\pi m$ and that the (proper) area is $16\pi m^2$.*

2.1.8 *Birkoff's theorem*

The coefficients of the Schwarzschild metric do not depend on t; such a metric is said to be *stationary*. Furthermore, there are no cross term involving spacelike and timelike increments (such as $dt d\phi$) in the metric. A stationary metric satisfying this further condition is said to be *static*. We have seen that the Einstein equations, which led from the general form (2.1) to the particular form (2.4) or (2.8), imply that a spherically symmetric vacuum solution is static. This result is known as Birkoff's theorem. It means that if the Sun were to oscillate while remaining spherical the surrounding vacuum would still be described by the Schwarzschild metric (2.4) and therefore that the oscillation could not be detected through the gravitational field. This accords with Newtonian gravity where a spherical distribution of matter gives the same gravitational field as a point of the same mass at its centre. One can emphasise the status of the theorem by rephrasing it as a uniqueness result: the Schwarzschild geometry is the unique spherically symmetric vacuum gravitational field satisfying Einstein's equations.

2.1.9 *Israel's theorem*

A converse to Birkoff's theorem can also be shown to hold with some additional technical assumptions: a static vacuum gravitational field must be spherical in general relativity, hence must be describable by the metric (2.4). We shall explore this in more detail later. Stated as a uniqueness theorem this becomes: the Schwarzschild geometry is the unique static vacuum gravitational field in general relativity. One is always free to choose different coordinates in relativity, so this result does not say that any spherically symmetric vacuum solution must look like (2.8)! But any such solution can be put into the form (2.8) by a suitable transformation of coordinates.

2.2 Orbits in Newtonian gravity

The simplest way to derive the equations of a test body orbiting in the gravitational field of a spherical mass M in Newtonian theory is to use the conservation of energy and angular momentum. (The energy and angular momentum are constants of the motion.) Since the orbit must lie in a plane (by conservation of angular momentum, or, more directly, by symmetry) we choose the equatorial plane, $\theta = \pi/2$, of a spherical polar coordinate system for convenience.

2.2.1 *Newtonian energy*

For a particle of unit mass in orbit in the equatorial plane the conservation of energy gives

$$E_{\mathrm{N}} = \frac{1}{2}\left(\left(\frac{dr}{dt}\right)^2 + r^2\left(\frac{d\phi}{dt}\right)^2\right) - \frac{GM}{r}, \tag{2.9}$$

where $E_{\mathrm{N}} = $ constant is the total energy per unit mass comprising the kinetic energy of radial and circular motion and the gravitational potential energy.

2.2.2 *Angular momentum*

Conservation of angular momentum implies (i) that the orbit lies in a plane and (ii) that the magnitude of the angular momentum per unit mass

$$L_{\mathrm{N}} = r^2\frac{d\phi}{dt} \tag{2.10}$$

is constant.

2.2.3 *The Newtonian effective potential*

Putting together (2.9) and (2.10) we get

$$\frac{1}{2}\left(\frac{dr}{dt}\right)^2 = E_{\mathrm{N}} - \frac{L_{\mathrm{N}}^2}{2r^2} + \frac{GM}{r} = E_{\mathrm{N}} - V_{\mathrm{N}}, \tag{2.11}$$

where V_{N} is called the effective potential. (It is *not* the usual gravitational potential itself because it contains the additional centrifugal term. In fact, it is the potential in a frame rotating with the orbit.)

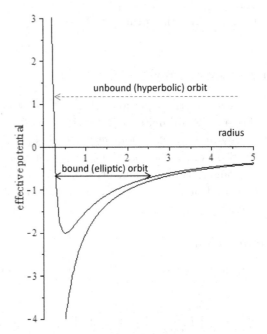

Fig. 2.1 Examples of the Newtonian effective potential as a function of radius for zero angular momentum (lower curve) and non-zero angular momentum (upper curve).

2.2.4 *Classification of Newtonian orbits*

Figure 2.1 shows a graph of the Newtonian effective potential as a function of radius r in the two cases $L_N = 0$ and $L_N = 1$ for a fixed central mass m. In interpreting the figure remember a horizontal line in the figure represents the motion of a particle (having constant total energy) and that the difference between the total energy and a given curve is the kinetic energy as a function of radius for given angular momentum: there is no implication that the motion is radial!

- For $L_N \neq 0$ we see that $(\frac{dr}{dt})^2$ is positive if $E_N > V_{\min}$. Thus the radial velocity is non-zero in these regions.
- From the figure we can see that orbits with $E_N > 0$ extend to arbitrarily large values of r. These are therefore the unbounded (hyperbolic) orbits.
- Orbits with $E_N < 0$ (but $> V_{\min}$) exist between fixed values of r; these are the bound (elliptic) orbits.

- There is one special case when $E_N = V_{min}$ where the value of r is fixed at $r_{circ} = L_N^2/GM$, which is clearly a circular orbit.

Note that, for a central point mass, if $L_N > 0$ a particle with any positive value of energy will be deflected by the potential back to infinity.

2.3 Particle orbits in the Schwarzschild metric

We are now ready to consider the paths of bodies which are subject to no non-gravitational forces, i.e. the trajectories in spacetime followed by bodies in free fall. These are given by geodesics of the metric (2.4). There are several ways to obtain such particle paths. The simplest is to make use of the symmetries to derive constants of the motion.

2.3.1 *Constants of the motion*

Let k^μ be a vector in a direction of symmetry, for example $k^\mu = (1, 0, 0, 0)$ along the t axis. Let $u^\mu = dx^\mu/d\tau$ be the tangent vector to the particle path $x^\mu = x^\mu(\tau)$ in spacetime and consider the component of k^μ along the path, namely $u_\mu k^\mu$. The rate of change of $u_\mu k^\mu$ in the direction u^ν is

$$u^\nu(k^\mu u_\mu)_{;\nu} = u^\nu u_{\mu;\nu}k^\mu + u^\nu u^\mu k_{\mu;\nu}$$

$$= u^\nu u_{\mu;\nu}k^\mu + \frac{1}{2}u^\nu u^\mu(k_{\mu;\nu} + k_{\nu;\mu}).$$

Suppose now that the path is a geodesic. Then $u^\nu u_{\mu;\nu} = 0$ because this is the geodesic equation. Furthermore, it can be shown that $k_{\mu;\nu} + k_{\nu;\mu} = 0$ for a vector along a symmetry direction (see Chapter 1). Thus $u^\nu(k^\mu u_\mu)_{;\nu} = 0$ and therefore $k^\mu u_\mu$ is constant along a geodesic (because, in words, the equation says that the derivative of $k^\mu u_\mu$ in the direction of the tangent vanishes).

2.3.2 *Conserved energy*

Consider first the time direction, $k^\mu = (1, 0, 0, 0)$. This is fairly obviously a symmetry of the static metric, since the metric coefficients are independent of t. (More formally, the metric is unchanged by the transformation $x^\mu \rightarrow x^\mu + \varepsilon k^\mu$.) We therefore have

$$g_{\mu\nu}k^\mu u^\nu = g_{00}k^0 u^0 = g_{00}u^0 = E = \text{constant}, \tag{2.12}$$

because the term in $k^0 = 1$ provides the only non-zero contribution to the sum. Note that we write this expression in terms of u^μ not u_μ because we know that $u^\mu = dx^\mu/d\tau$. Note that $u_0 = g_{0\nu}u^\nu = g_{00}u^0$, so (2.12) is equivalent to

$$u_0 = E \qquad (2.13)$$

(or $u_0 = E/c$ in physical units). Hence, from equation (2.12), along any geodesic in the Schwarzschild geometry, we have explicitly

$$(1 - 2m/r)\frac{dt}{d\tau} = E. \qquad (2.14)$$

The tangent vector u^μ to a geodesic is the 4-velocity vector of a particle in free-fall that follows the geodesic. Equivalently, u^μ is the momentum 4-vector of a unit mass particle. To find the meaning of the constant E consider the particle at infinity, then in this limit (2.14) becomes

$$\frac{dt}{d\tau} = E.$$

Now recall from Section 1.3.7 that the scalar product $u_{\text{obs}}^\mu u_\mu$ is the energy of a particle having 4-velocity u^μ in the frame of a local observer whose 4-velocity is u_{obs}^μ. For a stationary observer at infinity $(u_{\text{obs}}^\mu) = (1, 0, 0, 0,)$ so for a particle at infinity

$$u_{\text{obs}}^\mu u_\mu = \frac{dt}{d\tau} = E.$$

Thus E is the relativistic energy per unit mass of the particle relative to a stationary observer at infinity. The observer at infinity – sometimes called the bookkeeper – is an important notion. This observer uses the global Schwarzschild coordinates (r, θ, ϕ) and coordinate time t, the time kept by clocks at infinity. For this observer the energy of a free particle is conserved.

2.3.3 *Angular momentum*

Similarly, for the symmetry in the ϕ direction, we have $k^\mu = (0, 0, 0, 1)$ from which,

$$g_{\mu\nu}k^\mu u^\nu = g_{33}k^3 u^3 = g_{33}u^3 = -L = \text{ constant.} \qquad (2.15)$$

Note that $u_3 = g_{3\nu}u^\nu = g_{33}u^3$ so (2.15) is equivalent to

$$u_3 = -L. \qquad (2.16)$$

Using $u^3 = d\phi/d\tau$ and the Schwarzschild metric coefficient $g_{33} = -r^2$ this is explicitly

$$r^2\frac{d\phi}{d\tau} = L = \text{ constant.} \qquad (2.17)$$

In this case we interpret conservation of the ϕ component of momentum as the conservation of angular momentum, with L the angular momentum per unit mass relative to an observer at infinity. This interpretation is clearly borne out by comparison with the equivalent Newtonian equation. (Replace proper time τ by Newtonian time t and note that $r \times r\dot{\phi}$ is the angular momentum per unit mass.) This explains the choice of minus sign in (2.16).

Since the gravitational field is spherically symmetric, any free-fall orbit must lie in a plane. (Because all directions are equivalent, there is no way to choose a particular direction out of the plane in which to move. This is not true if the body is acted on by forces, such as rocket motors, because these can define a preferred direction.) As in the Newtonian case we take $\theta = \text{constant} = \pi/2$, the equatorial plane, so $\sin\theta = 1$, which turns out to be the most convenient choice.

Note that we prefer to use energy and angular moment *per unit mass* and hence 4-velocity rather than 4-momentum for particles, since particles of different mass follow the same worldline. We use 4-momentum whenever we want to compare the behaviour of a particle as it tends towards the speed of light with that of a photon.

2.3.4 *The effective potential*

Our final equation to determine the r coordinate along a geodesic can be obtained, most easily, from the scalar product $g^{\mu\nu} u_\mu u_\nu = 1$. Expanding and substituting for the metric coefficients and the conserved quantities u_0 and u_3, given by equations (2.13) and (2.16), gives

$$\left(1 - \frac{2m}{r}\right)^{-1} E^2 - \left(1 - \frac{2m}{r}\right) u_1^2 - \frac{L^2}{r^2} = 1.$$

Now, $u_1 = g_{11} u^1 = (1 - 2m/r)^{-1} (dr/d\tau)$, so finally we get

$$\left(\frac{dr}{d\tau}\right)^2 = E^2 - \left(1 + \frac{L^2}{r^2}\right)\left(1 - \frac{2m}{r}\right) = E^2 - V_{eff}. \qquad (2.18)$$

The quantity $V_{eff}(r)$, defined by this equation, is known as the effective potential. In physical units it has the dimensions of energy per unit mass, as does E^2/c^2. Writing the effective potential in inverse powers of r gives

$$V_{eff} = 1 - \frac{2m}{r} + \frac{L^2}{r^2} - \frac{2mL^2}{r^3}. \qquad (2.19)$$

From equation (2.19) we can deduce the shape of the potential. At infinity $V_{eff} = 1$; the second term on the right is an attractive potential that dominates at large r and pulls the potential downwards, until the third term, which is a repulsive centrifugal term, starts to dominate at smaller r, pulling the potential up to create a minimum. These terms are present in Newtonian gravity. The effect of the centrifugal term means that a particle with non-zero angular momentum, and energy $E > 1$, is reflected back to infinity by a Newtonian point mass (in a hyperbolic orbit). The final term in equation (2.19) is specific to relativistic gravity and starts to dominate the centrifugal term at small r, pulling the potential down and creating a maximum (Figure 2.2). This means that particles with angular momentum can be captured by a black hole provided that the particle energy exceeds the peak of the potential.

Henceforth we shall denote the mass of a test particle by m_0. This should not be confused with the geometric mass m of a black hole. Bringing together the results of the last sections we see that equations (2.14), (2.17) and (2.18) (and hence (2.19)) determine particle geodesics in the Schwarzschild spacetime. We shall now look at some applications of these equations.

Fig. 2.2 Examples of the relativistic effective potential, plotted as $[V_{\mathit{eff}} - 1]$ as a function of radius for $L/m = 4.3, 3.75, 3.46$.

2.3.5 *Newtonian limit*

We are now going to show how the Newtonian equations of motion for a body moving slowly (with respect to light) in an equatorial orbit can be obtained from an approximation to the Schwarzschild metric. Putting $\theta = \pi/2$ and dividing through the Schwarzschild metric by dt^2 we get

$$\left(\frac{d\tau}{dt}\right)^2 = 1 - \frac{2m}{r} - \left(1 - \frac{2m}{r}\right)^{-1}\left(\frac{dr}{dt}\right)^2 - r^2\left(\frac{d\phi}{dt}\right)^2. \qquad (2.20)$$

In the Newtonian limit velocities are small, so we can neglect the factor $(1 - 2m/r)^{-1}$ which multiplies $(dr/dt)^2$. (This factor represents a small correction to an already small quantity.) We then set $E^2 \approx 1 + 2E_N$, where E_N is small. (In physical units this is $E^2/c^2 \approx c^2 + 2E_N$.) We also need to use here the relation (2.14) for $d\tau/dt$ which, to the same order of approximation, is

$$\left(\frac{d\tau}{dt}\right)^2 \approx \frac{\left(1 - \frac{2m}{r}\right)^2}{1 + 2E_N} \approx \left(1 - \frac{4m}{r}\right)(1 - 2E_N) \approx 1 - \frac{4m}{r} - 2E_N,$$

neglecting products of small terms. With these approximations (2.20) gives the Newtonian equation (2.11). Since E_N is the Newtonian energy (per unit mass), E is clearly the relativistic energy (per unit mass) for a slowly moving particle.

We see that the metric

$$d\tau^2 = \left(1 - \frac{2m}{r}\right)dt^2 - dr^2 - r^2(d\theta^2 + \sin^2\theta d\phi^2) \qquad (2.21)$$

reproduces the Newtonian equations of motion for slowly moving bodies. We can therefore regard this as the Newtonian approximation to the metric (even though Newtonian gravity is not a metric theory, so does not strictly correspond to a spacetime structure with any metric). We see that in this approximation the gravitational potential produces a redshift, but not a curvature of space. This metric cannot account correctly for the perihelion precession of Mercury or for the bending of light.

Problem 2.2. *Show that at large radial distances and for low energies* $E^2/c^2 \approx c^2 + 2E_N$.

Problem 2.3. *Verify directly from the geodesic equations* (1.6) *that the Newtonian metric* (2.21) *reproduces the Newtonian equations of motion for* *slowly moving bodies. (You will need to compute the components of the affine connection as in Chapter 1 and neglect small quantities.)*

2.3.6 *Classification of orbits*

Figure 2.2 shows a plot of the effective potential $V_{\text{eff}}(r)$ against r. There are still regions of bound and unbound orbits, although there is no reason to believe these will be elliptical or hyperbolic now. Between the two there is a marginally bound (but not parabolic) orbit and the special case of a circular bound orbit. The most striking feature however, is that no amount of angular momentum can keep an orbit of sufficient energy out of the region $r < 2m$. Even worse, if $L < 2\sqrt{3}m$, all orbits end up inside $r = 2m$, since the centrifugal barrier vanishes below this value of L (see Fig. 2.2). These inward moving particles also have an azimuthal component of velocity, so they follow a spiral path into the horizon. Compare this with the Newtonian case (Fig. 2.1) where any angular momentum prevents a particle reaching $r = 0$. This is the first hint that if we can extend the vacuum field to such small radii then peculiar things will happen.

2.3.7 *Radial infall*

Let a particle be in radial free fall. Then its 4-velocity is $(u^\alpha) = (u^0, u^1, 0, 0)$ for purely radial motion. From (2.14) we have

$$u^0 = \frac{dt}{d\tau} = \frac{E}{\left(1 - \frac{2m}{r}\right)},$$

and, using $g_{\mu\nu}u^\mu u^\nu = 1$, we have

$$u^1 = \pm \left(E^2 - 1 + \frac{2m}{r}\right)^{1/2}. \tag{2.22}$$

The simplest orbit we can treat is the radial infall of a particle released from rest at infinity. In this case the particle at infinity has rest energy only, so $E = 1$. Hence $u^1 = \left(\frac{2m}{r}\right)^{1/2}$.

The velocity v of a particle falling from rest at infinity relative to an observer at infinity is $v =$ proper distance/coordinate time $= (1 - 2m/r)^{-1/2}dr/dt$. Now, $dr/dt = u^1/u^0 = (2m/r)^{1/2}(1 - 2m/r)$ so

$$v = \left(\frac{2m}{r}\right)^{\frac{1}{2}}\left(1 - \frac{2m}{r}\right)^{\frac{1}{2}}.$$

We see that at large r the velocity increases as r decreases in good agreement with Newtonian physics, but as the horizon is approached the velocity of infall slows to zero: as far as the distant observer is concerned the particle never crosses the horizon. We shall see later (Section 2.6.4), that the particle crosses the horizon in a finite lapse of its own proper time.

Problem 2.4. *Obtain the result* (2.22).

Problem 2.5. *Show that the escape velocity from radius r measured by a stationary observer at r is $\left(\frac{2m}{r}\right)^{1/2}$. What is the escape velocity from the horizon at $r = 2m$?*

2.3.8 *The locally measured energy of a particle*

We have just seen that for a distant observer the total energy per unit mass of the infalling particle has a constant value E. For comparison, the energy $\mathcal{E}$ (not per unit mass) measured by a local observer maintained at fixed coordinate position r with 4-velocity $(u^\mu) = (dt/d\tau, 0, 0, 0)$ is given by $u^\mu_{\text{obs}} p_\mu$, the projection of the particle 4-momentum along u^μ_{obs}.

For an observer at rest in the Schwarzschild spacetime, from the metric,

$$d\tau^2 = (1 - 2m/r)dt^2,$$

so

$$(u^\mu_{\text{obs}}) = ([1 - 2m/r]^{-1/2}, 0, 0, 0)$$

and

$$\mathcal{E} = u^\mu_{\text{obs}} p_\mu = \frac{m_0 E}{\left(1 - \frac{2m}{r}\right)^{1/2}}. \qquad (2.23)$$

It follows that as measured by local stationary observers the energy of the freely falling particle increases with decreasing r.

How is this compatible with the view of distant observers, who assign a fixed value $m_0 E$ to the energy? Equation (2.23) shows that the two energies are related by a redshift factor. It is helpful to understand why this is just what is required by energy conservation. Suppose the infalling particle is stopped by a stationary observer, its energy $\mathcal{E}$ converted to radiation and sent back to infinity. The energy of this radiation certainly suffers a redshift, so the energy received at infinity is $\mathcal{E} (1 - 2m/r)^{1/2}$. Unless this is equal to the energy $m_0 E$ that we started with, we should be able to construct a perpetual motion machine. Thus, (2.23) exactly expresses energy conservation.

We can obtain one more result from this argument. The kinetic energy of the particle in the local frames of reference is obtained by subtracting the rest mass energy from the total energy:

$$\mathcal{E} - m_0,$$

where as usual we put $c = 1$. Suppose that the particle is brought to rest at r and just the kinetic energy sent back to infinity as radiation. This energy suffers a redshift relative to the distant observer, so the energy received at infinity is

$$(\mathcal{E} - m_0) \left(1 - \frac{2m}{r}\right)^{1/2} = m_0 E - m_0 \left(1 - \frac{2m}{r}\right)^{1/2}.$$

Thus, according to the distant observer, after being brought to rest the particle is left with an energy

$$m_0 \left(1 - \frac{2m}{r}\right)^{1/2}, \tag{2.24}$$

this being the difference between the initial energy at infinity $m_0 E$ and the amount received back in radiation. As $r \to 2m$ the energy sent to infinity is just the original total energy $m_0 E$. The particle cannot give up any more energy to infinity than this. For suppose that its rest mass is converted to radiation at the horizon and sent back to infinity. According to (2.24) the energy received at infinity is redshifted to zero.

2.3.9 *Circular orbits*

Bringing an infalling particle to rest at a given radius, as in the previous section, is somewhat artificial. More realistically we can consider the circular orbits that are possible about the central mass.

For a circular orbit ($r = \text{constant}$) the two conditions

$$\frac{dr}{d\tau} = 0 \quad \text{and} \quad \frac{d^2 r}{d\tau^2} = 0 \tag{2.25}$$

must be satisfied. The first condition, in conjunction with equation (2.18), tells us that $E^2 = V_{\text{eff}}$, so the energy lies on the effective potential curve. The radial equation (2.18) gives also, in general, by differentiation with respect to τ,

$$\frac{d^2r}{d\tau^2} = -\frac{d}{dr}V_{\text{eff}}(r).$$

Hence, the second condition of (2.25) gives

$$\frac{dV_{\text{eff}}}{dr} = 0.$$

Therefore circular orbits are possible only at turning points of the effective potential.

Using the explicit form for $V_{\text{eff}}(r)$ from (2.18) we find the turning points occur at

$$r = \frac{L^2}{2m} \pm \frac{1}{2}\sqrt{\frac{L^4}{m^2} - 12L^2}. \tag{2.26}$$

For $L^2 > 12m^2$ there are two solutions, the negative sign in (2.26) corresponding to a maximum of $V_{\text{eff}}(r)$, which is an unstable point, and the positive sign to a minimum of $V_{\text{eff}}(r)$, which is therefore stable.

At $L^2 = 12m^2$ there is just one circular orbit: the maximum and minimum of the curve come together at a point of inflection for this value of L. This orbit is marginally stable ($d^2V_{\text{eff}}/dr^2 = 0$) and is the innermost (marginally) stable orbit. From (2.26) the condition $L^2 = 12m^2$ gives a radius $r = 6m$ for this innermost stable orbit. Because the orbit is only marginally stable, a particle at $r = 6m$ perturbed inward by a small amount will fall towards $r = 2m$. (At this stage we cannot say what will happen after that.)

2.3.10 *Comparison with Newtonian orbits*

In Newtonian gravity circular orbits can exist at any radius, however small, and are always bound. The angular momentum per unit mass of a particle in an orbit at r is $L = (GMr)^{1/2}$ and its binding energy per unit mass is $E = -GM/2r$. A particle moving in from infinity must lose energy before it can go into orbit, and an infinite amount of energy can be extracted from a particle falling into a 'Newtonian' black hole.

For a Schwarzschild black hole, stable circular orbits exist for $r \geq 6m$ only, and they are all bound. The angular momentum per unit mass of a particle in a circular orbit at radius r is

$$L = \left(\frac{mr}{1 - 3m/r} \right)^{1/2}$$

and the energy per unit mass is

$$E = \frac{1 - 2m/r}{(1 - 3m/r)^{1/2}}. \tag{2.27}$$

For $r < 6m$ circular orbits are unstable. In the range $4m < r < 6m$ the unstable orbits are bound ($E < 1$). At $r = 4m$ we have $E = 1$ and $L = 4m$. In the range $3m \leq r \leq 4m$ the orbits are unstable and unbound ($E \geq 1$). Thus, a particle coming in from infinity with $E \geq 1$ can (in principle) settle into an unstable circular orbit without losing any energy, provided that it has the appropriate value of angular momentum. Of course, since the orbit is unstable, in practice a small perturbation would eject the particle again. Nevertheless, this is a non-Newtonian feature of the strong gravitational field.

As $r \to 3m$ both E and L tend to infinity, so only zero mass particles (photons) can orbit at this radius.

Problem 2.6. *Show that the specific angular momentum of a particle in a circular orbit is given by*

$$L = \left(\frac{mr}{1 - 3m/r} \right)^{1/2}.$$

Show that the energy of this particle is

$$E = \left(1 - \frac{2m}{r} \right) \left(1 - \frac{3m}{r} \right)^{-1/2}.$$

Problem 2.7. *Show that the proper period of the orbit is*

$$\tau = 2\pi \left(\frac{r^3}{m} \right)^{1/2} \left(1 - \frac{3m}{r} \right)^{1/2}$$

and that the coordinate period is

$$T = 2\pi \left(\frac{r^3}{m} \right)^{1/2}.$$

Problem 2.8. *Show that there exists an unstable circular orbit into which a particle coming from infinity with specific energy $E = 1$ and impact parameter $L/E = 4m$ is inserted.*

2.3.11 Orbital velocity in the frame of a hovering observer

What is the orbital speed of a particle as measured by a hovering observer? We can calculate this as follows. The hovering observer has velocity 4-vector

$$(u_H^\mu) = \left((1 - 2m/r)^{-1/2}, 0, 0, 0\right),$$

and the orbiting particle has 4-velocity

$$(u_\mu) = (E, 0, 0, L),$$

where E is, as usual, the conserved energy per unit mass at infinity. Let the energy per unit mass of the particle be γc^2 with respect to the hovering observer. Then, with $c = 1$, the projection of the momentum of the particle on the 4-velocity of the observer gives

$$\gamma = u_\mu u_H^\mu = \frac{E}{(1 - 2m/r)^{1/2}}.$$

But recall from special relativity that, in terms of the velocity v of the orbiting particle (and with $c = 1$), $\gamma = (1 - v^2)^{-1/2}$. So using (2.27) for E and solving for v we find

$$v = \frac{(m/r)^{1/2}}{(1 - 2m/r)^{1/2}}.$$

Problem 2.9. *Show that the local orbital velocity in a circular orbit at radius r equals the escape velocity, $v_{\rm esc} = (2m/r)^{1/2}$, when $r = 4m$ and exceeds the escape velocity for $r < 4m$. Hence deduce that a circular orbit at $r < 4m$ is unbound.*

2.3.12 Energy in the last stable orbit

It is of interest to find the energy of a particle in the innermost stable orbit at $r = 6m$, because this will tell us how much energy can be extracted, in principle, from a particle spiralling slowly inwards. For a circular orbit, we have from (2.18)

$$E^2 = \left(1 - \frac{2m}{r}\right)\left(1 + \frac{L^2}{r^2}\right).$$

With $L^2 = 12m^2$ and $r = 6m$ this becomes

$$E^2 = 8/9.$$

Now a particle at rest at infinity has $E = 1$ (again from (2.18)). So the binding energy per unit mass of a particle in the innermost stable orbit is

$$1 - \sqrt{\frac{8}{9}} = 0.0572,$$

or about 5.7% of $m_0 c^2$ for a particle of mass m_0.

This means that matter spiralling inwards in, for example, an accretion disc (Section 6.8), will release 5.7% of its rest mass energy before it plunges into the hole from the last stable orbit. For comparison note that nuclear fusion yields about 0.7% of rest mass energy.

2.4 Orbits of light rays

The behaviour of light can also be used to show the effect of the strong gravitational field at small radii. To obtain the trajectories of light rays consider again the equations of motion for a particle of mass m_0. For conservation of energy we have from (2.14)

$$m_0(1 - 2m/r)\frac{dt}{d\tau} = m_0 E.$$

For a photon we let $d\tau \to 0$ and $m_0 \to 0$ with $d\lambda = d\tau/m_0$ finite. Then for a ray of light

$$(1 - 2m/r)\frac{dt}{d\lambda} = E_{ph}, \tag{2.28}$$

where $m_0 E \to E_{ph}$ as $m_0 \to 0$ and $E \to \infty$. Similarly, from equation (2.17), conservation of angular momentum gives

$$r^2\frac{d\phi}{d\lambda} = L_{ph}. \tag{2.29}$$

where $m_0 L \to L_{ph}$. In addition, we now have

$$d\tau^2 = 0,$$

or, equivalently $g^{\mu\nu}p_\mu p_\nu = 0$, where p^μ is the momentum 4-vector of a photon, from which

$$\left(\frac{dr}{d\lambda}\right)^2 = E_{ph}^2 - \frac{L_{ph}^2}{r^2}\left(1 - \frac{2m}{r}\right) = E_{ph}^2 - V_{ph}^2, \qquad (2.30)$$

where λ is an affine parameter along a ray. The corresponding effective potential $V_{ph}(r)$ has a maximum at $r = 3m$, which is independent of L_{ph}. This is therefore a circular orbit for photons, but because V_{ph} is a maximum the orbit is unstable. The potential $V_{ph}(r)$ has no minima, so there are no stable circular orbits for photons. In other words, although in principle the gravity of a spherical mass can act on light to keep it moving in a circle, in practice this orbit cannot be maintained.

Drawing together the above results, the equations (2.28), (2.29) and (2.30) govern the behaviour of light rays in the Schwarzschild geometry.

Problem 2.10. *Show that the orbital period, as measured by a distant observer, of a photon in a circular orbit about a spherical black hole is $c/\sqrt{3}$ and that the factor of $\sqrt{3}$ is a consequence of gravitational redshift.*

Problem 2.11. *Evaluate $d\tau/m_0$ in an inertial frame and show that in the limit as $d\tau \to 0$ and $m_0 \to 0$, $d\tau/m_0 \to dt/E_{ph}$, which is finite.*

Problem 2.12. *The local speed of light in relativity is always c. Verify that this is true for the Schwarzschild metric. (Hint: obtain an expression for the local velocity of a finite mass particle and consider the limit as its rest mass tends to 0.)*

2.4.1 Radial propagation of light

The 4-momentum of a light ray propagating radially is $(p^\mu) = (dt/d\lambda, \pm dr/d\lambda, 0, 0)$. Thus, from (2.28)

$$p^0 = \frac{E_{ph}}{\left(1 - \frac{2m}{r}\right)}$$

where E_{ph} is the conserved photon energy in the Schwarzschild frame. Now from $g_{\mu\nu}p^\mu p^\nu = 0$, we obtain $p^1 = \pm E_{ph}$. So

$$(p^\mu) = \left(\frac{E_{ph}}{\left(1 - \frac{2m}{r}\right)}, \pm E_{ph}, 0, 0\right).$$

We have seen that in Section 2.1.4 that light of wavelength λ sent from a stationary source at coordinate r to an observer at infinity suffers a redshift

$$\lambda_\infty = \lambda \left(1 - \frac{2m}{r}\right)^{-\frac{1}{2}}.$$

Now we consider the case of a probe in radial free-fall transmitting signals back to a spaceship at infinity. We want to find the relationship between the received and transmitted wavelengths. This can be done by evaluating the scalar product $u^\mu p_\mu$, where u^μ is the 4-velocity of the probe and p^μ is the 4-momentum of a photon. Evaluating the scalar product in a locally freely falling frame in which the probe is stationary gives

$$u^\mu p_\mu = h\nu,$$

where ν is the frequency of the signal in the rest frame of the probe. Now we evaluate the scalar product in the Schwarzschild frame, which is the frame of an observer at infinity. The components of the 4-velocity of the probe in the Schwarzschild frame, assuming it has dropped from rest at infinity, are

$$(u^\mu) = \left(\left(1 - \frac{2m}{r}\right)^{-1}, -\left(\frac{2m}{r}\right)^{\frac{1}{2}}, 0, 0\right),$$

where u^0 is given by equation (2.14) with $E = 1$. The u^1 component is obtained from $g_{\mu\nu}u^\mu u^\nu = 1$ with the assumption that the probe falls from rest at infinity. The covariant components of the outwardly propagating photon are

$$(p_\mu) = \left(h\nu_\infty, -\frac{h\nu_\infty}{\left(1 - \frac{2m}{r}\right)}, 0, 0\right).$$

So

$$u^\mu p_\mu = \frac{h\nu_\infty}{\left(1 - \frac{2m}{r}\right)} + \left(\frac{2m}{r}\right)^{\frac{1}{2}} \frac{h\nu_\infty}{\left(1 - \frac{2m}{r}\right)}$$

$$= h\nu_\infty \frac{\left[1 + \left(\frac{2m}{r}\right)^{\frac{1}{2}}\right]}{\left(1 - \frac{2m}{r}\right)}$$

$$= h\nu_\infty \left(1 - \left(\frac{2m}{r}\right)^{\frac{1}{2}}\right)^{-1} = h\nu,$$

where the final follows from the invariance of the scalar product. So finally we get for the wavelength at infinity

$$\lambda_\infty = \lambda \left(1 - \left(\frac{2m}{r} \right)^{1/2} \right)^{-1}. \tag{2.31}$$

Problem 2.13. *A spaceship plunges from rest at infinity into a black hole. Show that the frequency at which signals sent from infinity with frequency ν_∞ are received by the spaceship at the horizon is $\nu_\infty/2$. What is the frequency of this signal seen by the crew of the spaceship from inside the horizon (assuming that they survive for long enough)?*

2.4.2 Capture cross-section for light

The bending of light by gravity means that non-radial light rays can be captured by a black hole. Later we shall discuss the interior spacetime of a black hole $r < 2m$. For this we shall need to know which light rays can enter this region, that is, we shall need the capture cross-section of the black hole region for light rays. To look at the general properties of the orbits of light rays we return to the corresponding effective potential (Fig. 2.3)

$$V_{\text{eff}}^2 = \frac{L_{ph}^2}{r^2} \left(1 - \frac{2m}{r} \right).$$

The height of the maximum at $r = 3m$ is $L_{ph}^2/27m^2 = V_{\text{max}}^2$. For $E_{ph} > V_{\text{max}}$ an incoming photon enters $r = 2m$ so it is captured by the hole. For $E_{ph} < V_{\text{max}}$ an incoming photon is scattered by the potential back to infinity, so is not captured. The condition of capture is therefore $L_{ph}/E_{ph} < \sqrt{27}m$.

If the impact parameter of the photon that is just captured by the hole is b, then the capture cross-section of the hole is πb^2. So we have to relate the geometrical quantity b to the dynamical properties of the orbit, L_{ph} and E_{ph}. Now, the angular momentum of a particle at infinity is just the product of its 3-momentum, p, and its impact parameter b: $L_{ph} = pb$. For a photon we have $E_{ph} = p$ (recall that $c = 1$ in our units), so we can put $b = L_{ph}/E_{ph}$. This gives us the capture cross-section

$$\sigma = \pi b^2 = 27\pi m^2,$$

whereas the geometrical cross-section is $\pi r_s^2 = 4\pi m^2$.

Problem 2.14. *Using equation (2.30) and the conditions for a circular orbit (Section 2.3.9) show that a light ray coming in from infinity with*

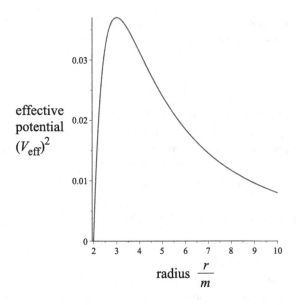

Fig. 2.3   Example of the relativistic effective potential for light rays.

impact parameter $b = 3\sqrt{3}m$ *enters an unstable circular orbit of radius* $r = 3m$.

It follows from problem 2.14 that a light ray coming from the unstable circular orbit at $r = 3m$ and travelling outwards will have an impact parameter $3\sqrt{3}m$. Thus to a distant observer the apparent radius of the black hole is $3\sqrt{3}m$. So, given a telescope with sufficient resolution, a Schwarzschild black hole will be seen as a black disc silhouetted against the stellar background. It should be possible to resolve the shadow cast by the supermassive black hole at the centre of our Galaxy using sufficiently long baseline interferometry at wavelengths $\lesssim 1\,\mathrm{mm}$ (see problem 2.15). We shall see (Section 3.10.3) that the shape of the shadow cast by a spinning black hole will not be circular, but will depend on the spin parameter of the black hole and the viewing angle relative to the rotation axis. The size and shape of the shadow can be used to determine the mass and spin parameters of a black hole.

Problem 2.15. *The black hole at the centre of our Galaxy has a mass of* $\sim 4 \times 10^6 M\odot$. *What is its apparent angular diameter as seen from the Earth, a distance of 8 kpc from the Galactic centre? This is of interest because it is now possible to resolve structures of the order of 40 micro arc* *seconds using very long baseline interferometry at wavelengths of the order of* 1.3*mm* (*Doeleman et al.*, 2008).

2.4.3 *The view of the sky for a stationary observer*

Suppose that a stationary observer at radius r looking at the sky in the equatorial plane $\theta = \pi/2$ receives a ray of light making an angle ψ' with the outward radial direction. This ray undergoes a proper radial displacement $dr/(1 - \frac{2m}{r})^{1/2}$ and a proper tangential displacement $rd\phi$ for each increment $d\lambda$ of affine parameter. From Fig. 2.4 the angle ψ' is therefore given by

$$\sin^2 \psi' = \frac{r^2 d\phi^2}{r^2 d\phi^2 + dr^2 / \left(1 - \frac{2m}{r}\right)}$$

$$= \frac{\left(1 - \frac{2m}{r}\right)}{\left(1 - \frac{2m}{r}\right) + \frac{1}{r^2}\left(\frac{dr}{d\phi}\right)^2}.$$

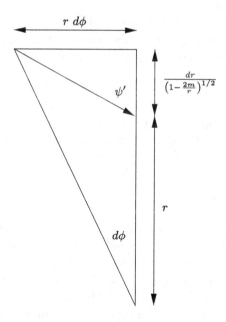

Fig. 2.4   Geometry of a light ray.

But $dr/d\phi = (dr/d\lambda)/(d\phi/d\lambda) = (dr/d\lambda)/(L_{ph}/r^2)$, and therefore from (2.30)

$$\left(\frac{dr}{d\phi}\right)^2 = \frac{E_{ph}^2}{L_{ph}^2}r^4 - \left(1 - \frac{2m}{r}\right)r^2,$$

and finally

$$\sin\psi' = \pm\left(1 - \frac{2m}{r}\right)^{1/2}\frac{b}{r} \qquad (2.32)$$

where $b = L_{ph}/E_{ph}$ is the impact parameter. We take the positive square root so that ψ' is near 0 for small b.

Equation (2.32) connects the angle of an incoming light ray measured by a local stationary observer at r to the impact parameter b of the ray. We can calculate the angle at which the outer edge of the black hole is seen by putting $b = \sqrt{27}m$ into equation (2.32). (See Section 2.4.2.) From (2.32) we see that as the observer hovers closer to the black hole, i.e. at smaller r coordinate, the angle ψ' gets smaller, so more of the sky is obscured by the black hole. At $r = 3m$, $\psi' = \pi/2$, so half the sky is black to the observer. In the limit as $r \to 2m$, $\sin\psi' \to 0$ and the area of the sky not covered by the black hole shrinks to zero.

Now, according to the calculation of the previous section, an observer close to the horizon at $r = 2m$ will receive only those rays with an impact parameter $b \le \sqrt{27}m$. The corresponding maximum angle is given by

$$\sin\psi'_{\max} = \left(1 - \frac{2m}{r}\right)^{1/2}\frac{\sqrt{27}}{2}$$

as $r \to 2m$. Thus the black hole fills the whole sky except for a disc of angular diameter $\sqrt{27}(1 - \frac{2m}{r})^{1/2}$, which closes up to zero for the observer at the horizon.

2.5 Classical tests

We have considered the special cases of radial and circular orbits of particles; now we look at more general orbits. The orbit equations (2.17) and (2.18) allow us to find $dr/d\phi$ as a function of r and hence to compute the paths of bound particles in space. The results can be compared with observation, for example with the perihelion precession of the planet Mercury. Similarly the equations of light rays allow calculation of the bending of light by, for example, the Sun. In these and other ways relativity can be compared with observation. In all of these tests the general relativistic result is found to be at the centre of ever-narrowing error bars. This is discussed further in standard texts on relativity.

These solar system tests involve weak gravitational fields so do not directly imply that strong field effects are correctly predicted by relativity. One example of a strong field test is the speeding up in the period of the binary pulsar (Taylor, 1994). But we also look for direct or indirect observation of black holes, having the predicted properties, to verify the theory. (See Chapter 5.)

Problem 2.16. *Obtain the equation of the orbit of a test mass with specific angular momentum L in the equatorial plane of a spherical mass m in the weak field limit $GM/rc^2 \ll 1$*

$$\frac{d^2u}{d\phi^2} + u = \frac{m}{L^2} + 3mu^2,$$

where $u = 1/r$. By seeking an approximate solution of the form $u = l^{-1}(1 + \varepsilon \cos \lambda \phi)$ for small eccentricity ε, show that the perihelion rotates by $6\pi m/l$ per orbit.

Obtain similarly the equation of motion of a light ray

$$\frac{d^2u}{d\phi^2} + u = 3mu^2.$$

Show that in the absence of the non-Newtonian term on the right of this equation the motion of light is a straight line. With this term, $3mu^2$, included, verify that there is a unique circular orbit for photons.

2.6 Falling into a black hole

From an astrophysical point of view we are interested in the collapse of a sphere of matter that cannot support itself against gravity. This was first discussed by Oppenheimer and Snyder (1939). From Birkoff's theorem on the uniqueness of the Schwarzschild metric we already know that the collapse of a spherical star from the point of view of an external observer is described by the Schwarzschild geometry. To be precise let us assume that the pressure support in the star has been removed completely and that the surface of the star is in free fall. This gives us a simple way of discussing the approach of the surface towards $r = 2m$. Specifically, we want to know how much time it takes, according to an observer located at a large fixed value of r, for the surface to approach $r = 2m$.

2.6.1 *Free-fall time for a distant observer*

For radial free fall we can refer to Section 2.3.7 to obtain (2.33) below, or we start again from (2.18),

$$\left(\frac{dr}{d\tau}\right)^2 = E^2 - \left(1 - \frac{2m}{r}\right),$$

where we have put $L = 0$ for radial infall. Assuming that the collapse started with a negligible inward speed,

$$\frac{dr}{d\tau} \to 0 \text{ as } r \to \infty,$$

so $E^2 = 1$. Thus

$$\left(\frac{dr}{d\tau}\right)^2 = \frac{2m}{r}. \tag{2.33}$$

However, viewing the collapse from a large distance, we do not record the proper time of the infalling matter, but we use our own proper time, which is the coordinate time t. Thus we use

$$\frac{dt}{d\tau} = \frac{1}{1 - \frac{2m}{r}},$$

from which

$$\frac{dr}{dt} = -\left(\frac{2m}{r}\right)^{1/2}\left(1 - \frac{2m}{r}\right) \tag{2.34}$$

describes the infall of the surface as seen from a distance. We integrate this from a large value of r, $r = R$, say, to close to $r = 2m$ (using r' as the integration variable to distinguish it from the infall radius):

$$t = -\int_R^r \left(\frac{r'}{2m}\right)^{1/2} \frac{r'dr'}{r' - 2m}.$$

To approximate the integral note that the major contribution is from close to $r' = 2m$ so we put $r' = 2m + \varepsilon$ and hence, expanding in powers of ε,

$$\left(\frac{r'}{2m}\right)^{1/2}\frac{r'}{r'-2m} = \frac{2m}{\varepsilon} + \cdots,$$

where the higher order terms not shown are small compared to $1/\varepsilon$. Hence,

$$t \sim -2m \int_R^r \frac{dr'}{r' - 2m},$$

approximately, and

$$t \sim -2m \log\left(\frac{r}{2m} - 1\right),$$

the contribution from the lower limit $r' = R$, which we have neglected, being much smaller than the term we retain provided that r is close enough to $2m$. This is the time taken for the collapsing surface to approach $r = 2m$ from a large radius. Notice that it tends to infinity as $r \to 2m$.

2.6.2 *Light-travel time*

But the observer will not see the approach until a photon from the surface has had time to reach him. Therefore, we have to add the time taken by a ray of light to reach a large distance from close to $r = 2m$.

For a light ray $d\tau^2 = 0$ so, from the metric with $d\theta = d\phi = 0$ also

$$\frac{dt}{dr} = \frac{1}{1 - \frac{2m}{r}}.$$

Integrating

$$t' = \int_r^R \frac{r' dr'}{r' - 2m} \sim -2m \log\left(\frac{r}{2m} - 1\right),$$

again taking the main contribution to the integral to come from close to $r = 2m$.

The observer sees the surface at time $T = t + t'$ where, therefore,

$$T = -4m \log\left(\frac{r}{2m} - 1\right). \tag{2.35}$$

This tends to infinity at $r = 2m$. Thus, for an external observer the infall to $r = 2m$ takes an infinite time.

2.6.3 *What the external observer sees*

The wavelength λ_∞ of radiation received at large r coming from a freely falling source is

$$\lambda_\infty = \lambda \left(1 - \left(\frac{2m}{r} \right)^{\frac{1}{2}} \right)^{-1}.$$

See Section 2.4.1. So the redshift is

$$1 + z = \frac{\lambda_\infty}{\lambda} = \left(1 - \left(\frac{2m}{r} \right)^{\frac{1}{2}} \right)^{-1} = \left(\frac{2m}{r} \right)^{-\frac{1}{2}} \left(\left(\frac{r}{2m} \right)^{\frac{1}{2}} - 1 \right)^{-1}.$$

To express this in terms of the time T of a distant observer rewrite equation (2.35) as follows:

$$\left(\left(\frac{r}{2m} \right)^{\frac{1}{2}} + 1 \right) \left(\left(\frac{r}{2m} \right)^{\frac{1}{2}} - 1 \right) = \exp\left(-\frac{T}{4m} \right).$$

Hence

$$1 + z = \left(\frac{2m}{r} \right)^{-\frac{1}{2}} \left(\left(\frac{r}{2m} \right)^{\frac{1}{2}} + 1 \right) \exp\left(\frac{T}{4m} \right).$$

So as $r \to 2m$ the external observer sees the redshift tend to infinity exponentially quickly as the surface of the star approaches $r = 2m$.

Similarly the luminosity decreases proportionally to $1/(1 + z)^2$ (one factor of $1 + z$ coming from the redshift and one from time dilation: the photon energy falls and so does the rate of arrival of photons). Thus an external observer sees the luminosity of the star decline in time as $\exp(-T/2m)$ as it collapses to $r = 2m$. (A more accurate calculation would take into account non-radial rays leading to a luminosity that falls approximately as $\exp(-T/3\sqrt{3}m)$.) In physical units the decay time is $\sim 2GM/c^3$ which is around a microsecond for a solar mass black hole!

Seen by a distant observer the coordinate velocity dr/dt of an infalling object increases during the initial stages of the fall but later, as it approaches the horizon, it decreases to hover forever just outside the horizon. This assumes the object is viewed through infinitely sensitive instruments that can continue to detect it despite its exponentially fading luminosity. Thus, from the point of view of an external observer a collapsing star would never form a black hole. This caused a lot of confusion in the 1960s and led Russian astronomers to call black holes 'frozen stars'. That black holes do form can be seen by looking at the collapse in the reference frame of the falling object.

2.6.4 *An infalling observer's time*

From equation (2.33) in terms of the proper time of an infalling observer we have

$$\tau = -\int_R^{2m} \left(\frac{r}{2m}\right)^{1/2} dr = \frac{4}{3}m\left(\left(\frac{R}{2m}\right)^{3/2} - 1\right)$$

as the time to fall from $r = R$ to $r = 2m$. Thus the infalling observer reaches $r = 2m$ from a finite distance in a finite proper time.

A falling object can also be viewed by observers situated on static shells that it passes on its way. The velocity of the object relative to the shell at coordinate r is given by

$$v_{\text{loc}} = \frac{\text{proper distance}}{\text{proper time}} = \frac{dr}{(1 - 2m/r)dt}.$$

Using (2.34) for an object falling from rest, we get

$$v_{\text{loc}} = \left(\frac{2m}{r}\right)^{1/2}. \tag{2.36}$$

(An equivalent result was obtained in Section 2.3.7.) Thus the locally measured velocity of infall increases steadily with decreasing r and in the limit as $r \to 2m$ we get $v_{\text{loc}} \to 1$ (i.e. in physical units the locally measured speed tends to c).

Problem 2.17. a) *A particle falls from rest at infinity. Show that the time it takes to fall from $r = 2m$ to $r = 0$ is $4m/3$.*
b) *Show that the maximum time that a particle can take in going from $r = 2m$ to $r = 0$ is $m\pi$. (Hint: use the metric and the condition that the world line of the falling particle must be light-like in the limit.)*
c) *Show that it corresponds to a particle falling from rest at $r = 2m$.*

2.6.5 *What the infalling observer feels*

Seen from a local frame fixed with respect to the Schwarzschild coordinates a radially infalling particle having $E = 1$ falls with speed $v_{\text{loc}} = (2m/r)^{1/2}$ (Section 2.6.1, or equation (2.36)). If we work from the frame of the freely falling particle, the speed of the coordinate lattice in this frame is also

$v_{\mathrm{loc}} = (2m/r)^{1/2}$. Let τ be the proper time of the particle. The acceleration of the stationary frame is $dv_{\mathrm{loc}}/d\tau = (dv_{\mathrm{loc}}/dr)(dr/d\tau) = (dv_{\mathrm{loc}}/dr)$ $v_{\mathrm{loc}} = m/r^2$.

As in Newtonian physics the dependence of the acceleration on radial distance means that an extended body will feel a tidal force. We want to calculate this tidal force from the point of view of the infalling observer (who will after all be subject to the force). We follow the derivation in Taylor and Wheeler (2000). Any two observers agree on their relative acceleration, so the relative acceleration of the head and the feet of a radially aligned infalling observer is

$$dg = \frac{2m}{r^3}dr.$$

To get the tidal force we need the proper distance corresponding to a coordinate distance dr (because our height that we measure in metres is our proper height in free fall, not, in principle, a coordinate displacement in Schwarzschild coordinates, although the two will turn out to be numerically the same).

In a stationary frame the proper distance corresponding to a displacement dr is $dr(1 - 2m/r)^{-1/2}$. For the freely falling observer moving with speed $v_{\mathrm{loc}} = (2m/r)^{1/2}$, this length is contracted by a factor

$$1/\gamma = (1 - v_{\mathrm{loc}}^2)^{1/2} = (1 - 2m/r)^{1/2}.$$

So the proper length in the freely falling frame is

$$\gamma^{-1}dr(1 - 2m/r)^{-1/2} = dr.$$

The tidal acceleration on a body of extension Δr is therefore approximately $(2m/r^3)\Delta r$. At some point this will begin to cause pain for the infalling observer. We want to calculate how long this pain must be endured before the singularity at $r = 0$ is encountered.

Suppose that our criterion for the onset of pain is $\Delta g = g_{\mathrm{E}}$, the gravity at the Earth's surface. This will occur at a radius $r_p = (2m\Delta r/g_{\mathrm{E}})^{1/3}$. The infall time to $r = 0$ is obtained by integrating (2.33):

$$\tau = -\int_{r_p}^{0} \left(\frac{r}{2m}\right)^{1/2} dr = \frac{2r_p^{3/2}}{3(2m)^{1/2}} = \frac{2}{3}\left(\frac{\Delta r}{g_{\mathrm{E}}}\right)^{1/2}.$$

Note that this is independent of the black hole mass.

Problem 2.18. *Show that for a standard observer of height $2m$ the infall time from the threshold of pain to death at the singularity is about 0.3s.*

Show that for an observer falling into a $1M_\odot$ *black hole the threshold of pain occurs well outside the Schwarzschild radius. For what mass black hole does the pain threshold coincide with the Schwarzschild radius?*

Problem 2.19. *What mass of black hole would allow an observer in free fall to survive for a year inside the horizon before being destroyed?*

2.7 Capture by a black hole

In Section 2.4.2 we looked at the probability that a ray of light would be captured by a black hole taking into account the bending of light by the gravitational field of the black hole. It is of interest to calculate the corresponding capture cross-section of a black hole for particles. The method is similar: any particle with energy $E > V_{\text{eff}}^{\max}$ is captured by the hole. This gives us a critical angular momentum for capture. Instead of angular momentum (for a unit mass particle) L we use the impact parameter b defined in terms of the particle velocity at infinity v by $L = vb$. The capture cross-section is πb^2. With a lot of algebra it can be shown that the maximum of

$$V_{\text{eff}}^2 = \left(1 - \frac{2m}{r}\right)\left(1 + \frac{L^2}{r^2}\right)$$

occurs for $r = (L^2/2m)\left[1 - \left(1 - 12m^2/L^2\right)^{1/2}\right]$ and is given by

$$(V_{\text{eff}}^{\max})^2 = \frac{1}{54}\left[l^2 + 36 + (l^2 - 12)\left(1 - \frac{12}{l^2}\right)^{1/2}\right], \tag{2.37}$$

where $l = L/m$.

For the rest of the calculation we consider two limiting cases.

2.7.1 *Case I: Capture of high angular momentum particles*

Inserting the condition $l \gg 1$ into (2.37) and using the binomial expansion gives

$$(V_{\text{eff}}^{\max})^2 \approx \frac{l^2 + 9}{27}.$$

The critical angular momentum below which capture will occur is therefore $l_*^2 = 27E^2 - 9$. Thus, for a particle of energy E per unit mass and momentum

p per unit mass, the critical impact parameter is

$$b_* = \frac{L_*}{p} = \frac{ml_*}{(E^2-1)^{1/2}},$$

and the cross-section is therefore

$$\sigma = \pi b_*^2 = \frac{\pi m^2 l_*^2}{(E^2-1)}.$$

For large E ($E \gg 1$) we have $(E^2-1)^{-1} = E^{-2}(1-1/E^2)^{-1} \approx E^{-2}(1+1/E^2)$, so

$$\sigma \approx \frac{\pi m^2}{E^2}\left(1+\frac{1}{E^2}\right)(27E^2-9) = 27\pi m^2\left(1+\frac{2}{3E^2}\right).$$

Note that as $E \to \infty$, $\sigma \to 27\pi m^2$, and we recover the capture cross-section for a photon.

2.7.2 *Case II: Capture of low energy particles*

For a particle with low energy ($v \ll 1$) we put $(V_{\text{eff}}^{\max})^2 = E^2 \approx (1+\frac{1}{2}v^2)^2 \approx 1+v^2$ as the limit for capture and solve for the critical angular momentum l_* in terms of v. This gives

$$l^2 \approx 16 + 27v^2 + \cdots,$$

neglecting terms in v^4 and higher. Thus

$$\sigma = \frac{16\pi m^2}{v^2},$$

which agrees with our intuition that a particle moving slowly enough will be captured by the gravity of the black hole.

2.8 Surface gravity of a black hole

In this section we are going to derive an important quantity associated with a black hole, called its surface gravity by analogy with the acceleration due to gravity g of Newtonian theory. First we need to determine the acceleration of a particle at rest with respect to a local observer at rest, and then to determine how this acceleration appears from infinity.

2.8.1 *The proper acceleration of a hovering observer*

The proper or rest frame acceleration a of a particle is related to its 4-acceleration a^μ by equation (1.13):

$$a^\mu a_\mu = -a^2,$$

where

$$a^\mu = \frac{du^\mu}{d\tau} + \Gamma^\mu_{\rho\sigma} u^\rho u^\sigma.$$

The components of the 4-velocity of a hovering observer are

$$u^\mu = \left(\left(1 - \frac{2m}{r}\right)^{-\frac{1}{2}}, 0, 0, 0 \right).$$

The only non-zero component of the 4-acceleration is

$$a^1 = \frac{du^1}{d\tau} + \Gamma^1_{00} (u^0)^2$$

where

$$\Gamma^1_{00} = \frac{m}{r^2} \left(1 - \frac{2m}{r}\right).$$

As $u^1 = 0$ and the energy of the hovering observer at r is $E = (1-2m/r)^{1/2}$, we obtain

$$a^1 = \frac{m}{r^2}, \quad \text{and} \quad a_1 = g_{11} a^1.$$

Finally,

$$a^\mu a_\mu = -\left(\frac{m}{r^2}\right)^2 \left(1 - \frac{2m}{r}\right)^{-1},$$

so

$$a = \frac{m}{r^2} \left(1 - \frac{2m}{r}\right)^{-\frac{1}{2}}.$$

Note that the proper acceleration goes to infinity at the horizon, so only a photon can hover at the horizon.

2.8.2 *Surface gravity*

What is the gravitational acceleration at the event horizon as seen from infinity? The result in fact has the Newtonian form m/r^2. (Note that this is not the locally measured gravitational acceleration at the hole, which we have just calculated in the previous section, so it is not the same as Newtonian gravity, nor is it the Newtonian approximation at large distances that we are dealing with, even though this has the same form.) We shall give two ways of arriving at this answer, offering different levels of physical insight and mathematical sophistication. First, a simple physical approach.

Consider a massless inextensible string used by a distant observer to raise a particle of unit mass at a radius r through a distance dl. In this process the local energy of the particle increases by $dE_r = g_r dl$, where g_r is the acceleration due to gravity at r. The energy expended by the distant observer in this process is $dE_\infty = g_\infty dl$. By conservation of energy the two energies are related by a redshift factor. (Otherwise one could make a perpetual motion machine by converting energy to radiation and sending it between the two ends of the string.) Thus

$$\frac{g_r}{g_\infty} = \frac{E_r}{E_\infty} = \left(1 - \frac{2m}{r}\right)^{-1/2} \tag{2.38}$$

so

$$g_\infty = \frac{m}{r^2}. \tag{2.39}$$

Here g_r is the proper acceleration felt by an observer hovering at radius r, which would be interpreted as the acceleration due to gravity at the radius r. It is given by (2.8.1). As $r \to 2m$, $g_r \to \infty$. Nevertheless, the force required at infinity to hold the particle of unit mass hovering at the horizon is $g_\infty = m/(2m)^2 = 1/4m$. We call this the *surface gravity* κ of the black hole. Our result is therefore that the surface gravity of a Schwarzschild black hole is $\kappa = 1/4m$.

2.8.3 *Rindler coordinates*

To investigate further what the spacetime is like in the vicinity of the event horizon we introduce a new set of coordinates in this region. For the time coordinate we use the Schwarzschild t, but we define a new radial coordinate

ξ centred on $r = 2m$ and symmetrical about this origin by

$$r - 2m = \frac{\xi^2}{8m}$$

where the factor of 8 has been inserted with hindsight to tidy up the resulting expressions. We also define

$$\kappa = \frac{1}{4m},$$

which we have seen is the surface gravity of the black hole. Then we find

$$1 - \frac{2m}{r} = \frac{(\kappa\xi)^2}{(1 + \kappa^2\xi^2)} \tag{2.40}$$

$$\sim \kappa^2\xi^2 \tag{2.41}$$

near $\xi = 0$ (i.e. near $r = 2m$). So near the horizon the metric is approximately

$$d\tau^2 \sim \kappa^2\xi^2 dt^2 - d\xi^2 - \frac{1}{4\kappa^2}d\Omega^2, \tag{2.42}$$

where

$$d\Omega^2 = d\theta^2 + \sin^2\theta d\phi^2.$$

The coordinates (t, ξ, θ, ϕ) are called Rindler coordinates.

This metric has a simple interpretation. Consider just the (t, ξ) plane and let

$$T = \xi \sinh \kappa t \tag{2.43}$$

$$X = \xi \cosh \kappa t.$$

Substitution into the (t, ξ) part of the Rindler metric (2.42) shows that

$$d\tau^2 = dT^2 - dX^2, \tag{2.44}$$

which is just Minkowski spacetime in two dimensions.

Problem 2.20. *Derive equation* (2.44).

So for an observer hovering near the horizon the Rindler metric is just Minkowski spacetime in an unusual coordinate system! Near the horizon of a spherical black hole the geometry is approximately flat.

This may require some explanation. To remain near the horizon keeping r close to $2m$ an observer must apply an acceleration. (Un-accelerated trajectories are radial geodesics which fall into the hole.) It is along this accelerated trajectory that the metric is approximately flat. This is a special feature. In a general spacetime it is only (unaccelerated, freely falling) geodesic observers who see their neighbourhoods as approximately flat.

Problem 2.21. *The proper time interval given by the Rindler metric in two dimensions is*

$$d\tau^2 = \kappa^2 \xi^2 dt^2 - d\xi^2.$$

On the curve $\xi = 1/a = $ constant we have, from equation (2.43),

$$T = a^{-1} \sinh \kappa t,$$
$$X = a^{-1} \cosh \kappa t.$$

From this calculate the components of the proper acceleration on this trajectory in (T, X) coordinates, namely $d^2T/d\tau^2$ and $d^2X/d\tau^2$, and show that the magnitude of the proper acceleration is a.

Thus a is the proper acceleration on the curve $\xi = 1/a$. From equation (2.41) $a \sim \kappa\left(1 - \frac{2m}{r}\right)^{-1/2}$. Comparing this with (2.8.1) we see that κ is approximately the acceleration of a particle near the horizon as seen from infinity, a relation that becomes exact at the surface of the hole at $r = 2m$. The quantity κ is therefore equivalent to the Newtonian acceleration due to gravity ('g') which again justifies calling it the surface gravity of the hole.

2.9 Other coordinates

The Schwarzschild metric appears to be a viable metric, and a solution of Einstein's equations, for $r < 2m$. However, in this region, the coordinate r is timelike and the coordinate t spacelike! For example, for a body in the region $r < 2m$ with $\theta = $ constant, $\phi = $ constant, $t = $ constant we have $d\tau^2 = g_{11}dr^2 > 0$ so a (non-zero) dr is a timelike interval in this region. This means that a body in this region cannot remain at constant r (since dr must be non-zero to make the worldline of the body timelike). We shall see that it must fall into the singularity at $r = 0$. Furthermore, since the metric coefficients depend on time, the metric within $r = 2m$ is not static.

However, the connection between the space within $r = 2m$ and outside $r = 2m$ cannot be examined in Schwarzschild coordinates, because the metric is invalid at $r = 2m$. We say that there is a coordinate singularity at

$r = 2m$, because the problem turns out to be with the coordinate system, not with the physics. (A similar problem occurs at the pole of a spherical polar coordinate system in Euclidean space, where $\theta = 0$, $0 \le \phi \le 2\pi$ is not a circle, but a single point.) We can remedy this by finding one or more alternative sets of coordinates that do not have a coordinate singularity at $r = 2m$. We shall follow the historical development, introducing first the two sets of Eddington–Finkelstein coordinates and then the Kruskal coordinates, which cannot be expressed in elementary functions, but which have the advantage that they cover the whole spacetime.

2.9.1 *Null coordinates*

Consider first Minkowski spacetime. On an outgoing light ray (r increasing as t increases), $u = t - r$ is a constant. Different values of u label different light rays according to the initial point (the value of r when $t = 0$). As usual for convenience we use units with $c = 1$. Similarly, on ingoing rays the quantity $v = t + r$ is a constant, different values of the constant labelling different rays. Along an ingoing ray u changes, so u labels points on an ingoing ray. Conversely v labels points on a given outgoing ray. Because they label light rays the coordinates u and v are often called null coordinates. We can use either u or v (or both) as an alternative 'time' coordinate. For example, making the transformation

$$t = v - r \tag{2.45}$$

we get

$$d\tau^2 = dv^2 - 2dvdr - r^2 d\widetilde{\omega}^2.$$

Using u and v as coordinates in the (r, t) plane, we have

$$d\tau^2 = dudv - \frac{1}{4}(v - u)^2 d\widetilde{\omega}^2.$$

2.9.2 *Eddington–Finkelstein coordinates*

To carry out the analogous transformation in Schwarzschild spacetime we need to find the equation of a radial null ray. This is obtained by integrating $d\tau^2 = 0$ for a radial ray. We have from (2.8)

$$dt^2 = \frac{dr^2}{\left(1 - \frac{2m}{r}\right)^2} = dr_*^2, \tag{2.46}$$

where we have defined the new radial coordinate r_* in such a way that radial null rays are $d(t \pm r_*) = 0$ or $t \pm r_* =$ constant. Integrating the second pair of equations in equation (2.46) we get

$$r_* = r + 2m \log \left| \frac{r - 2m}{2m} \right|. \tag{2.47}$$

Let

$$v = t + r_*.$$

We are going to use v as a time coordinate, replacing t. Substitution in the Schwarzschild metric gives

$$d\tau^2 = \left(1 - \frac{2m}{r}\right)(dt^2 - dr_*^2) - r^2 d\widetilde{\omega}^2 \tag{2.48}$$

$$= \left(1 - \frac{2m}{r}\right) dv^2 - 2drdv - r^2 d\widetilde{\omega}^2. \tag{2.49}$$

The final expression is the form of the Schwarzschild metric in ingoing Eddington–Finkelstein coordinates. In this form the metric is non-singular at $r = 2m$, because $\det(g) = -(2m)^4 \sin^2\theta \neq 0$. To plot the spacetime in a (t, r) plane we introduce the new time coordinate corresponding to v and r,

$$t_* = v - r$$

(compare (2.45)). Figure 2.5 shows the (t_*, r) plane extending from $r = 0$ to $r = \infty$. In particular we can see what happens to the light cones as we approach $r = 2m$, which is a non-singular surface in these coordinates. Either from the figure, or from the metric (2.49), we see that $r = 2m, \theta = \text{constant}, \phi = \text{constant}$ is a null ray, and that the surface $r = 2m$ is generated by these null rays (for different values of θ and ϕ).

2.10 Inside the black hole

We are now in a position to consider the properties of the black hole inside the event horizon. We can show that within the surface $r = 2m$ all future pointing directions point inwards to decreasing r. This means it is impossible to escape from within the region $r < 2m$, since real bodies can travel only in timelike (or null) directions. The surface $r = 2m$ therefore acts as a one-way membrane for matter and radiation: nothing can escape into the surrounding spacetime. The surface $r = 2m$ is the event horizon, and the region $r < 2m$ is a 'black hole'.

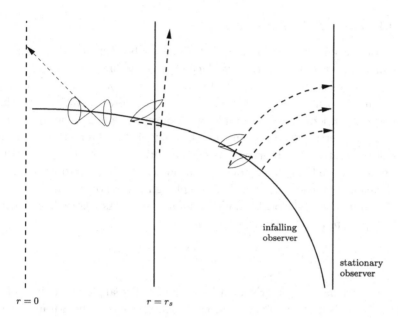

Fig. 2.5 Spacetime diagram in Eddington–Finkelstein coordinates showing an observer falling towards a black hole.

To demonstrate this take the metric (2.49) and write it as

$$2drdv = -\left[d\tau^2 - \left(1 - \frac{2m}{r}\right)dv^2 + r^2 d\widetilde{\omega}^2\right]$$
$$= -\left[d\tau^2 + \left(\frac{2m}{r} - 1\right)dv^2 + r^2 d\widetilde{\omega}^2\right].$$

For a timelike or null displacement $d\tau^2 \geq 0$ so for $r < 2m$ the sum of squares on the right-hand side is positive and the overall expression is negative. Recall now that on a radial null ray $dv = d(t + r_*) = d(t_* + r) = 0$, and hence that a displacement dv lies within the future light cone if $dv > 0$. So to make the left-hand side negative for a future-pointing displacement we have to have $dr < 0$. Thus a future-pointing displacement is necessarily in an inward direction.

The event horizon cuts off all events within $r = 2m$ from the outside world, so an explorer who crosses the event horizon of a black hole can never return or signal his experiences back to the outside world. A more pressing problem is that the observer cannot avoid destruction at the singularity that lies in their future.

2.10.1 *The infalling observer*

In Section 2.4.3 we showed that a stationary observer hovering close to the horizon sees an almost totally black sky: in the limit as r tends to $2m$ the whole sky becomes black. Recall that stationary observers do not exist inside the horizon so we cannot use them to explore its interior. In this section we will look at the example of an observer falling freely from rest at infinity. Aberration will mean that the view of this observer differs from that of the stationary observer outside the horizon. In addition we will be able to follow the freely falling observer inside the horizon up to the point of annihilation at the singularity. Let the plunging observer fall from rest at infinity and denote his local frame of reference by S. As before, his velocity with respect to a stationary observer is

$$v = \left(\frac{2m}{r}\right)^{1/2}. \tag{2.50}$$

Let the stationary observer's local frame be S'. The angle of the incoming light ray in this frame is ψ'; let the corresponding angle in the S-frame be ψ. The usual special relativistic velocity transformation applies between the local frames, so

$$\cos\psi = \frac{\cos\psi' - v}{1 - v\cos\psi'} \tag{2.51}$$

(recalling that we use $c = 1$). Thus $\cos\psi \langle\cos\psi'$ so $\psi\rangle\psi'$: the observer in S sees less of the sky as black.

From equation (2.32) for $\sin\psi'$ we get directly that

$$\cos\psi' = \left[1 - \left(1 - \frac{2m}{r}\right)\frac{b^2}{r^2}\right]^{1/2}.$$

With this expression for $\cos\psi'$, equation (2.51) becomes

$$\cos\psi = \frac{\left[1 - \left(1 - \frac{2m}{r}\right)\frac{b^2}{r^2}\right]^{1/2} - v}{1 - v\left[1 - \left(1 - \frac{2m}{r}\right)\frac{b^2}{r^2}\right]^{1/2}}. \tag{2.52}$$

The angle $\psi_{\max}$ at which the falling observer sees the edge of the black hole is obtained by setting $b = \sqrt{27}m$ into (2.52). (See Section 2.4.2.) From (2.52) we saw that the effect of aberration is to make this angle larger than the angle seen by the hovering observer. So now let us evaluate this angle at the point where the observer crosses the horizon. Inserting $r = 2m$ into (2.52) directly gives an indeterminate form. So we put $r = 2m + \delta$ and expand both numerator and denominator using the binomial theorem for small δ. In the limit as $\delta \to 0$ this gives

$$\cos \psi_{\text{max}} = \frac{1 - \frac{27}{4}}{1 + \frac{27}{4}} = -\frac{23}{31}$$

or $\psi_{\text{max}} = 138°$. Thus the black hole is seen in the direction of fall subtending an angle of $84°$.

The angle ψ varies smoothly as the horizon is crossed so evidently we can use equation (2.52) inside the horizon. So let us work out what the plunging observer sees at $r = m$. Equation (2.52) together with (2.50) and $b = \sqrt{27}m$ gives $\cos \psi_{\text{max}} = -0.6$, or $\psi_{\text{max}} = 127°$, so the angle subtended by the black hole is $106°$. Thus, on moving inwards, the observer sees more of the sky to be black. Finally as the observer approaches the singularity, i.e. as $r \to 0$, equation (2.52) becomes

$$\cos \psi_{\text{max}} \sim -\left(\frac{r}{2m}\right)^{1/2} \tag{2.53}$$

which is independent of the impact parameter b. Hence, in the limit, all rays are seen at $\cos \psi = 0$, or $\psi_{\text{max}} = \pi/2$ and the black hole subtends an angle of $180°$. So as the observer crashes into the singularity the outside world appears as a thin bright ring.

Problem 2.22. *Derive* (2.53) *from equation* (2.52).

2.11 White holes

The Eddington–Finkelstein coordinates we used above were based on inward going null rays that entered $r = 2m$ in the future. It is natural to ask if we can make a similar picture with outgoing rays by taking $u = t - r_*$ as the 'time' coordinate. Straightforward substitution into the Schwarzschild metric gives

$$d\tau^2 = \left(1 - \frac{2m}{r}\right) du^2 + 2drdu - r^2 d\widetilde{\omega}^2.$$

This describes (part of) a spacetime which is the time-reverse of the black hole in which all future-directed null or timelike paths emerge from $r < 2m$ into the surrounding space. This object is called a white hole. Note that both black holes and white holes are mathematical solutions of the vacuum Einstein equations but that does not mean they are both physical possibilities. We shall present arguments later (Chapter 7) to support the existence of black holes, but there is no evidence for the existence of white holes.

2.12 Kruskal coordinates

The Eddington–Finkelstein coordinates cover either the part of spacetime representing a black hole region or the part representing a white hole, but not both. It would obviously be convenient to be able to look at both regions in the same picture. This is made possible by yet another coordinate system, the Kruskal coordinates.

Return to the Schwarzschild metric in the form

$$d\tau^2 = \left(1 - \frac{2m}{r}\right)(dt^2 - dr_*^2) - r^2 d\Omega^2$$

and write this in double null coordinates u and v as

$$d\tau^2 = \left(1 - \frac{2m}{r}\right) du\, dv - r^2 d\Omega^2.$$

Now define

$$U = -\exp(-u/4m) \tag{2.54}$$

$$V = \exp(v/4m). \tag{2.55}$$

This can be shown to lead to (problem 2.23)

$$d\tau^2 = \frac{32m^3}{r}\exp(-r/2m)dU\,dV - r^2 d\tilde{\omega}^2, \tag{2.56}$$

where r is considered to be a function of U and V through

$$VU = -\exp\left(\frac{v - u}{4m}\right) = -\exp(r_*/2m),$$

since $v - u = 2r_*$. The Kruskal coordinates are (U, V, θ, ϕ). In Kruskal coordinates light rays are represented by lines at $45°$ (with $c = 1$).

The metric (2.56) covers four regions as shown in Fig. 2.6. The two regions $r > 0$ and $r < 0$ are separate asymptotically flat universes. Note the two singularities at $r = 0$, one in the past and one in the future, which appeared separately in the pictures of the two sets of Eddington–Finkelstein coordinates. Figure 2.8 shows the surface of a collapsing star within which the metric would differ from the Kruskal picture.

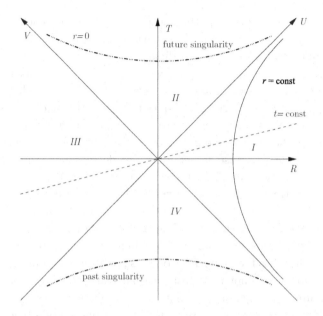

Fig. 2.6   Schwarzschild spacetime in Kruskal (U, V) coordinates.

There is one aspect of the Kruskal picture that can be misleading. Each point in the plane represents a 2-sphere of that radius at that time. This is not the same as saying that either of the angles can be re-created by rotating the diagram about an axis (which is how one often re-introduces an additional dimension). It is absolutely not the case that one can get from region I to region II, even in principle, by moving on a circle of constant ϕ! (And conversely, of course, rotation in Euclidean space in spherical polar coordinates does not take one to a region of negative r.)

Problem 2.23. *Obtain the Kruskal metric from the coordinate transformations (2.54) and show that the light cones in the metric are everywhere upright.*

2.12.1   *The singularities at $r = 0$ and cosmic censorship*

After falling through the event horizon a particle must continue to move to smaller values of r. This can continue in principle until the particle approaches $r = 0$ where the curvature becomes infinite. Any real body will suffer large tidal distortions as it approaches $r = 0$ and will be destroyed.

But what happens to our hypothetical point particle? It reaches the edge of spacetime beyond which the theory has nothing to say. This is a true physical singularity, unlike the coordinate singularity at $r = 2m$.

To a certain extent it is a harmless singularity as far as physics in the outside world is concerned. This is because the unpredictability is hidden behind an event horizon and does not affect the black hole exterior. The white hole singularity at $r = 0$ in the past is a different matter. Here any unpredictability emerges into the surrounding spacetime making physics impossible. (Because, for example, any detection of a particle can be attributed to its emergence from the singularity rather than having to have a prior cause within the physical universe.) Thus the white hole singularity is 'naked'.

We shall see further examples of naked singularities later. The cosmic censorship hypothesis (put forward by Penrose) seeks to make naked singularities illegal as far as 'normal' physics is concerned by proposing that naked singularities cannot form from normal matter. There is no proof of this conjecture, but attempts to violate it have only helped to frame it more precisely. For the present, we observe that the formation of a black hole by the collapse of matter is not a time-symmetric process and leads to a black hole without the associated white hole.

2.12.2 *The spacetime of a collapsing star*

Figures 2.7 and 2.8 show how the collapse of a sphere of matter to a singularity at $r = 0$ can be represented in (a) Eddington–Finkelstein coordinates and (b) Kruskal coordinates respectively. Note that within $r = 2m$ the Eddington–Finkelstein space and time coordinates (t, r_*) swap roles so the singularity at $r_* = r = 0$ occurs at a constant 'time' and is therefore spacelike, despite what the picture appears to show at first glance! This is clearly represented in the Kruskal picture. Note also that the white hole region has disappeared to be replaced by a solution of the field equations for the interior of the star. Although we make no attempt to find an interior solution, the pictures are drawn under the (correct) assumption that, prior to the final collapse, the interior will be a regular spacetime with a *decreasing* curvature as we go through the star towards $r = 0$.

Fig. 2.7 The spacetime of a collapsing star in Eddington–Finkelstein coordinates.

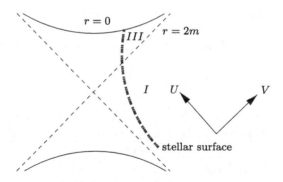

Fig. 2.8 The spacetime of a collapsing star in Kruskal coordinates.

2.13 Gullstrand–Painlevé coordinates

The Gullstrand–Painlevé coordinate system has a pedagogical advantage: it describes the principal features of the Schwarzschild geometry by means of a simple analogy, the flow of a fluid, a picture that can be easily visualised. We obtain the Gullstrand–Painlevé metric from the Schwarzschild form as follows.

We imagine a stationary observer on a spherical shell at coordinate radius r. The velocity of a particle falling freely from rest at infinity with respect to this observer is $v = -(2m/r)^{1/2}$ (Section 2.3.7). A time interval dt_s measured by the stationary observer at r is related to the Schwarzschild coordinate time t by

$$dt_s = \left(1 - \frac{2m}{r}\right)^{\frac{1}{2}} dt \qquad (2.57)$$

The proper distance between adjacent shells, as measured by the stationary observer, is

$$dr_s = \frac{dr}{\left(1 - \frac{2m}{r}\right)^{\frac{1}{2}}} \qquad (2.58)$$

from the Schwarzschild metric. In the falling frame this distance is seen length-contracted so

$$dr_{FF} = \frac{dr_s}{\gamma} = \frac{dr}{\gamma(1 - \frac{2m}{r})^{\frac{1}{2}}}.$$

But $\gamma = (1 - v^2)^{-1/2} = (1 - 2m/r)^{-1/2}$ so

$$dr_{FF} = dr.$$

So the proper distance in the falling frame is equal to the coordinate distance in the Schwarzschild frame. Note that this is true only for a particle falling from rest at infinity.

Now we can relate the local frame of the falling observer to the local frame of the stationary observer at r by a Lorentz transformation. This gives:

$$dt_{FF} = \gamma(dt_s - v dr_s).$$

Using equations (2.57) and (2.58) we get

$$dt_{FF} = dt - v\gamma^2 dr.$$

Finally, substituting for dt in terms of dt_{FF} in the Schwarzschild metric, we get the Gullstrand–Painlevé metric

$$ds^2 = \left(1 - \frac{2m}{r}\right) dt^2_{FF} - 2 \left(\frac{2m}{r}\right)^{\frac{1}{2}} dr dt_{FF} - dr^2 - r^2 d\theta^2 - r^2 \sin^2 \theta d\phi^2.$$

$$(2.59)$$

Note that the coordinate singularity at $r = 2m$ has been removed. Note also that the metric is no longer static but stationary and is not time-symmetric, reflecting the one-way nature of the horizon. Of course, it shares these features with the Eddington–Finkelstein form.

The coordinate representation (2.59) of the Schwarzschild geometry lends itself to a simple picture of a black hole (Hamilton and Lisle, 2008). A particle at rest in the Gullstrand–Painlevé frame is being swept inwards towards the horizon like a swimmer in a river flowing towards a waterfall. This is, of course, how we set up the metric, but to see it directly we write the metric as

$$ds^2 = \left(1 - \frac{4m}{r}\right) dt^2_{FF} - \left(dr + \left(\frac{2m}{r}\right)^{\frac{1}{2}} dt_{FF}\right)^2 - r^2 d\theta^2 - r^2 \sin^2 \theta d\phi^2,$$

$$(2.60)$$

which gives us back a static (diagonal) form if $dr/dt_{FF} = -(2m/r)^{1/2}$. Upstream of the waterfall it is possible to swim against the current and to move upstream. But, at the top of the falls, the flow velocity equals the maximum speed of the swimmer (the speed of light). At this point, a swimmer endeavouring to swin upstream is stationary with respect to the shore, just as is a photon emitted outwards at the horizon of a black hole. Once over the top of the falls the speed of the river is greater than the maximum speed of the swimmer so she cannot avoid being swept down to destruction at the base of the falls.

Again we can see this from the metric by recasting it in the form

$$ds^2 = \left[dr + \left(1 + \left(\frac{2m}{r}\right)^{\frac{1}{2}}\right) dt_{FF}\right]$$

$$\times \left[dr - \left(1 + \left(\frac{2m}{r}\right)^{\frac{1}{2}}\right) dt_{FF}\right] - dr^2 - r^2 d\Omega^2.$$

For radial propagation of light $ds^2 = 0$ and the angular increments are zero, so we get

$$\frac{dr}{dt_{FF}} = \pm \left[1 - \left(\frac{2m}{r} \right)^{\frac{1}{2}} \right]$$

where the positive sign is for outward propagation and the negative sign for inward propagation. At large values of r, we have $dr/dt_{FF} \approx \pm 1$ and the light cones are upright. But as r is reduced, the light cones tip in the direction of decreasing r until at the horizon, $r = 2m$, the outward directed ray is stationary. For $r < 2m$, dr/dt_{FF} is always negative, and all particles, irrespective of whether they are outward or inward directed are swept ineluctably towards the singularity at $r = 0$. The river picture also enables us to appreciate how tidal forces arise. The feet of an infalling observer are moving faster that her head, so the observer is stretched. She is also squeezed because of the convergence of the flow.

Problem 2.24. *Use the Gullstrand–Painlevé metric to show that the redshift z of radiation sent from an observer at a radial point r to a distant stationary observer is given by*

$$1 + z = \left(1 - \frac{2m}{r} \right)^{-\frac{1}{2}}.$$

2.14 Embedding diagrams

The geometrical image of a cylinder as the surface of a tube is highly misleading in many respects. Intrinsically the cylinder has the geometry of flat space. The curvature one 'sees' when it is represented as a surface in three dimensional space comes from the embedding (the extrinsic geometry) of a 2-surface in three-dimensional space. Nevertheless, the embedding picture of the cylinder does help to inform our feeling for what the geometry is like. To get some feel for the global geometry of the Schwarzschild black hole we can try to represent aspects of it by embeddings in three-space in a similar way.

To do this of course, we have to suppress all but two dimensions. Let us look at the spaces of constant time and also suppress one of the angular coordinates, by putting, say, $\theta = \pi/2$. The metric is

$$dl^2 = \left(1 - \frac{2m}{r} \right)^{-1} dr^2 + r^2 d\phi^2.$$

The metric of the Euclidean embedding space is

$$dl^2 = dz^2 + dr^2 + r^2 d\phi^2,$$

which on $z = z(r)$ becomes

$$dl^2 = (1 + z'^2)dr^2 + r^2 d\phi^2,$$

where $z' = dz/dr$. These are the same if

$$z = 2(2m)^{1/2}(r - 2m)^{1/2}. \tag{2.61}$$

This gives the surface in Fig. 2.9.

This is the way in which black holes are customarily represented in popular literature. Since the Schwarzschild coordinates are not valid for $r = 2m$ we cannot follow the geometry further in by this means. If we look at the Kruskal picture however we see that through the bifurcation point at the origin the two Schwarzschild regions join together. For this cross-section (such as the one labelled $t = $ constant in Fig. 2.6) we get the Einstein–Rosen bridge as the embedding diagram, Fig. 2.10.

Although we cannot show it here, the only difference between this and a general section through the Kruskal metric, such as $(U + V) = $ const. in Fig. 2.6, is that the throat region is extended and narrower, and breaks into two separate pieces if the section intersects $r = 0$.

The two regions below and above the throat in Fig. 2.10 are two separate asymptotically flat universes. One might anticipate the possibility

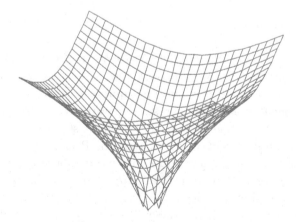

Fig. 2.9   A plot of equation (2.61) showing the embedding surface with the same geometry as Schwarzschild in the $r - \phi$ plane.

Fig. 2.10   The Einstein–Rosen bridge.

of travelling through the throat from one to the other. However, a glance at the full Kruskal picture shows that any attempt must end up in the singularity. In effect, the throat closes off faster than an observer can travel through it. On the other hand, it is not hard to imagine a picture in which the throat would connect two parts of the same universe, in which case we have a *wormhole*. (See Chapter 5.)

2.15 Asymptotic flatness

Far enough away from the central source of gravity the Schwarzschild spacetime approximates to Minkowski spacetime. Subject to certain technical definitions of what we mean by 'approximates to' such spacetimes are called *asymptotically flat*. For such spacetimes we can get what is sometimes a useful global picture of the causal structure (i.e. of which events can be causally connected) by using the Penrose–Carter diagrams. These make it possible to draw pictures of what is happening at infinity, by changing the coordinates to bring infinity to a finite coordinate value. This distortion is carried out in such a way that the relationship between light rays is maintained. We say that these diagrams show the causal structure of spacetimes (but clearly not their metric structure, since infinity is in the wrong place).

To see what this means consider the transformation from the Lorentzian coordinates (t, r, θ, ϕ) to new coordinates $(\psi, \xi, \theta, \phi)$ where

$$t = \frac{1}{2}\left(\tan\frac{1}{2}(\psi + \xi) + \tan\frac{1}{2}(\psi - \xi)\right) \qquad (2.62)$$

$$r = \frac{1}{2}\left(\tan\frac{1}{2}(\psi + \xi) - \tan\frac{1}{2}(\psi - \xi)\right).$$

The Minkowski metric becomes

$$d\tau^2 = \Omega^{-2}[d\psi^2 - d\xi^2 - \sin^2\xi(d\theta^2 + \sin^2\theta d\phi^2)] = \Omega^{-2}d\tilde{\tau}^2,$$

where $\Omega = 2\cos\frac{1}{2}(\psi + \xi)\cos\frac{1}{2}(\psi - \xi)$. Neglecting the overall factor Ω we get a new metric $d\tilde{\tau}^2$ which has the same light paths as $d\tau^2$ (because the vanishing of one implies and is implied by the vanishing of the other). The replacement of the metric $d\tau^2$ by the metric $d\tilde{\tau}^2$ is called a conformal transformation, with conformal factor Ω.

The geometry of $d\tilde{\tau}^2$ is shown in Fig. 2.11. Light rays begin in the past at $\mathfrak{I}^-$ and end at $\mathfrak{I}^+$ (the lines $\psi + \xi = \pi$ and $\psi + \xi = -\pi$). These boundaries therefore represent light-like infinity. Causal relationships are respected by the transformations. Nevertheless, the paths of massive

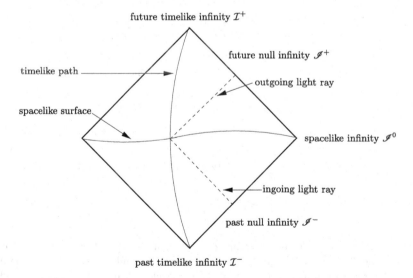

future timelike infinity $\mathcal{I}^+$

future null infinity $\mathscr{I}^+$

timelike path

outgoing light ray

spacelike surface

spacelike infinity $\mathscr{I}^0$

ingoing light ray

past null infinity $\mathscr{I}^-$

past timelike infinity $\mathcal{I}^-$

Fig. 2.11 Asymptotic structure of Minkowski spacetime shown in a Penrose–Carter diagram.

particles in the two metrics are not the same. All timelike paths in $d\tilde{\tau}^2$ start from $\mathcal{I}^-$ and end at $\mathcal{I}^+$. Arranging that a finite range $(-\pi, \pi)$ of the new coordinates ψ and ξ cover an infinite range of the original (t, r) coordinates in (2.62), and making the conformal transformation, enables us to form a finite picture of an infinite spacetime. Such pictures are called Penrose or Penrose–Carter diagrams. In the next section we consider the Penrose–Carter diagram for Schwarzschild spacetime.

2.15.1 *The Penrose–Carter diagram for the Schwarzschild metric*

Our extension of the spacetime has taken us from the exterior region covered by the Schwarzschild coordinates to the 'whole' spacetime covered by the Kruskal coordinates. Although we have not proved it, in fact no further extensions are possible. We can put this together with a further conformal transformation that has no effect near the horizon, but which brings infinity to a finite distance. This provides us with a Penrose–Carter diagram of the black hole and allows us to picture its causal structure (Fig. 2.12).

The figure gives a view of the (U, V) plane.

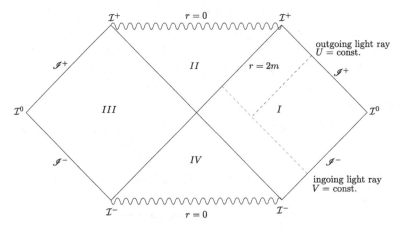

Fig. 2.12 Penrose–Carter diagram of the Kruskal spacetime showing the asymptotic structure. Outgoing light rays end at null infinity on $\mathcal{J}^+$ and incoming rays begin at null infinity on $\mathcal{J}^-$. Spacelike infinity is denoted I^0 and timelike future and past infinity $\mathcal{I}^+$ and $\mathcal{I}^-$ respectively. Light rays are given by $U = $ constant and $V = $ constant.

2.15.2 *The Penrose–Carter diagram for the Newtonian metric*

Out of interest we can give a similar treatment for the Newtonian metric (equation (2.21)). The picture at infinity is the same as that for flat spacetime, but we need new coordinates to look at $r = 2m$. Confining ourselves to the (t, r) plane we put

$$dr = \pm \left(1 - \frac{2m}{r}\right)^{1/2} dR.$$

Then

$$d\tau^2 = \left(1 - \frac{2m}{r}\right)(dt^2 - dR^2).$$

The metric $d\tilde{\tau}^2 = d\tau^2/(1 - 2m/r) = dt^2 - dR^2$ is therefore flat. Near $r = 2m$ we have

$$R \sim 2\sqrt{2m}(r - 2m)^{1/2},$$

so $r = 2m$ corresponds to $R = 0$. The metric is singular at $R = 0$, because $r < 2m$ clearly does not yield a Lorentzian metric. (Of course, near $r = 2m$ the Newtonian metric is not a physical gravitational field, because it is not even an approximate solution to Einstein's equations, but that does not stop us from considering the spacetime geometry to which it corresponds.)

2.16 Non-isolated black holes

There are several distinct concepts that coalesce in the Schwarzschild black hole, but which are different in general for black holes in a non-empty environment, for example for a black hole accreting material from the surrounding medium. In general we have to distinguish the following: the infinite redshift surface, the event horizon and the apparent horizon.

2.16.1 *The infinite redshift surface*

The surface defined by $g_{00} = 0$ has an infinite redshift to an external observer. To determine if a point belongs to the infinite redshift surface requires a knowledge of the metric just at that point, so this surface is a local construction. Thus it will not necessarily coincide with the event horizon in general.

2.16.2 *Trapped surfaces*

Consider a two-dimensional spacelike surface and imagine that it emits a flash of light. A lightfront moves normal to the surface inwards and outwards. Under normal circumstances the area of the inward moving lightfront decreases and that of the outward moving one increases. Equivalently, the inward normals to the inward moving lightfront converge and the outward normals to the outward moving front diverge.

However, if we repeat this experiment inside a black hole, both sets of normals converge (Fig. 2.13). Our two-dimensional spacelike surface is said to be trapped.

2.16.3 *Apparent horizon*

The outermost trapped surface is the apparent horizon. In the vacuum Schwarzschild solution this is the surface $r = 2m$. However, the apparent event horizon is clearly a local construction (you can measure the convergence or divergence locally), whereas the ('true') event horizon is a global property: it depends on constructing null geodesics to determine whether they reach infinity or not, which cannot be judged from the local behaviour.

For a concrete example consider a black hole accreting a spherical shell of matter (Frolov and Novikov, 1998). A light ray which was heading to

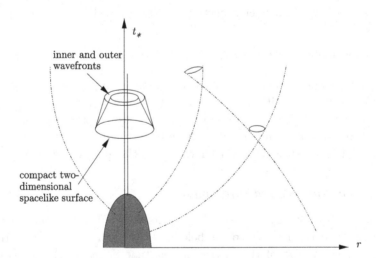

Fig. 2.13   Both outgoing and ingoing wavefronts from a trapped surface converge. The trapped surface is inside a black hole formed by collapse.

infinity before the infall of the shell is subject to a stronger gravitational field after the matter is accreted, so may not in fact escape to infinity. Thus the original ray is part of the event horizon even though it was not initially trapped. The apparent horizon therefore lies inside the event horizon. One can show that this is a general feature of the two horizons.

2.17 The membrane paradigm

Although the event horizon is not a physical boundary (and is locally undetectable) it is intuitively convenient to think of a black hole as a physical object, endowed with physical properties that encode its relation to the external world. To do this rigorously, we imagine the black hole surrounded by a membrane to which are attributed appropriate physical properties. This is the so-called 'membrane paradigm' (Thorne *et al.*, 1986). For example, to express the influence of a black hole on external electromagnetic fields the membrane is endowed with a surface resistivity allowing charges and currents to terminate electric and magnetic field lines. Similarly, to account for the deformations of a black hole in the presence of, for example, an orbiting planet, the membrane can be given certain elastic properties. Normal physical bodies have timelike surfaces. The membrane is therefore chosen not to be coincident with the horizon, which is a null surface, but as a timelike surface just outside the horizon. The black hole can then be thought of from the outside as a normal body in spacetime. This approach works not only for Schwarzschild black holes but also for the more general vacuum black holes of the next chapter, and also for black holes in general, which we do not consider here. The details are beyond the scope of this book but can be found in Thorne *et al.* (1986).

Chapter 3

Rotating Black Holes

We saw in the previous chapter that the Schwarzschild metric is the only spherically symmetric vacuum solution of Einstein's field equations and that this represented the unique end-point of the collapse of a spherical star. Since stars are not strictly spherical but axisymmetric about an axis of rotation, it is natural to seek axially symmetric vacuum solutions of Einstein's equations to represent the spacetime outside a rotating star. Such solutions can be found and may be of interest, for example, in the structure of rotating neutron stars. For a slowly rotating body of mass M with angular momentum J the Einstein equations outside the body are solved approximately by the metric

$$ds^2 \approx \left(1 - \frac{2GM}{c^2 r}\right) c^2 dt^2 + \frac{4GJ \sin^2 \theta}{c^2 r} dt d\phi$$

$$- \left(1 + \frac{2GM}{c^2 r}\right) dr^2 - r^2 d\theta^2 - r^2 \sin^2 \theta d\phi^2, \qquad (3.1)$$

to order $1/r$. As $J \to 0$ we recover the Schwarzschild solution.

Another way of approaching this question is to imagine a star acquiring angular momentum by capture of infalling, orbiting material. Outside the star the metric will be given by (3.1). Now imagine the capture of an infalling particle from the last stable orbit by a Schwarzschild black hole. We have already seen that all but 5.7% of the rest mass energy of the particle is added to the mass of the black hole. In addition, the black hole captures the angular momentum of the matter in the last stable orbit. This leads to a spinning black hole. One might imagine again that the metric would be given by (3.1). Far from a spinning black hole this is true; close to a spinning black hole it is not even approximately the case (because the $O(1/r^2)$ terms cannot be neglected), no matter how slowly the black hole is rotating. Instead we have a new solution, the Kerr solution, which gives the pure vacuum solution corresponding to a spinning black hole. Note that, conversely, the Kerr solution does not appear to correspond to the field outside a rotating material body, since no interior solutions with physical matter have been found that fit smoothly on to a Kerr exterior metric.

We shall next give the Kerr solution without proof. (Even to check by direct substitution that it satisfies the vacuum Einstein equations is a major undertaking!)

3.1 The Kerr metric

The following metric is a solution of the Einstein field equations in a vacuum:

$$ds^2 = \frac{(\Delta - a^2 \sin^2 \theta)}{\rho^2} c^2 dt^2 + \frac{4GMa}{c\rho^2} r \sin^2 \theta d\phi dt$$

$$- \frac{\rho^2}{\Delta} dr^2 - \rho^2 d\theta^2 - \frac{A \sin^2 \theta}{\rho^2} d\phi^2,$$

where

$$\Delta = r^2 - \frac{2GM}{c^2} r + a^2, \tag{3.2}$$

$$\rho^2 = r^2 + a^2 \cos^2 \theta, \tag{3.3}$$

$$A = (r^2 + a^2)^2 - a^2 \Delta \sin^2 \theta, \tag{3.4}$$

which we shall find to be the metric of a black hole of mass M and angular momentum $J = aMc$. This solution is the Kerr metric in Boyer–Lindquist coordinates (t, r, θ, ϕ). In these coordinates the metric is approximately Lorentzian at infinity (i.e. it is Minkowski space in the usual coordinates of special relativity). For $a = 0$ we recover the Schwarzschild solution in Schwarzschild coordinates.

Note that the parameter a has units of length. The metric can be simplified by defining time and masses in length units also. To do this we put: $m = GM/c^2$ and replace $s = c\tau$ by τ and ct by t as we did for the Schwarzschild metric. The result is equivalent to setting $G = 1$ and $c = 1$.

We obtain

$$d\tau^2 = \frac{(\Delta - a^2 \sin^2 \theta)}{\rho^2} dt^2 + \frac{4ma}{\rho^2} r \sin^2 \theta d\phi dt$$

$$- \frac{\rho^2}{\Delta} dr^2 - \rho^2 d\theta^2 - \frac{A \sin^2 \theta}{\rho^2} d\phi^2. \tag{3.5}$$

or, with a slight rearrangement, in an equivalent form

$$d\tau^2 = \frac{\Delta}{\rho^2}(dt - a \sin^2 \theta d\phi)^2 - \frac{\sin^2 \theta}{\rho^2}[(r^2 + a^2)d\phi - adt]^2 - \frac{\rho^2}{\Delta} dr^2 - \rho^2 d\theta^2, \tag{3.6}$$

where

$$\Delta = r^2 - 2mr + a^2, \tag{3.7}$$

and ρ and A are unchanged. For future reference we define the angular momentum of the black hole $J = aMc$ in physical units. In geometrical units we put $j = am = GJ/c^3$. The units of j are (metres)^2.

3.2 The event horizon

The first property of the Kerr metric that we notice is that as $\Delta \to 0$ the metric coefficient $g_{11} \to \infty$. So this appears to be the analogue of the Schwarzschild condition $r \to 2m$. Indeed, if $a = 0$, then $\Delta = 0$ when $r = 2m$. Anticipating somewhat, we see that $\Delta = r^2 + a^2 - 2mr = 0$ is a quadratic in r and so has two solutions $r_\pm$. It is indeed the case that the outer of these solutions,

$$r_+ = m + \sqrt{m^2 - a^2}, \tag{3.8}$$

is the location of the event horizon in the Kerr metric. The justification that r_+ is the event horizon in the equatorial plane will be provided in Section 3.10, and that it is the event horizon in general will be apparent from the Penrose–Carter diagram in Section 3.18. The quantity Δ is negative between r_- and r_+ so the r-coordinate is timelike in this region. The inner solution r_- has a different physical significance which we shall explain in Section 3.18. In the next section we shall evaluate two quantities associated with the event horizon to provide some practice with the Kerr metric and for future reference.

Problem 3.1. *Show that Δ is negative between r_- and r_+ and that r is timelike in this region.*

3.2.1 *The circumference of the event horizon*

The (proper) circumference of the curve $r = $ constant in the equatorial plane $\theta = \pi/2$ at constant Boyer–Lindquist time t, is

$$\int_0^{2\pi} (-g_{33})^{1/2} d\phi = \int_0^{2\pi} \frac{[(r^2 + a^2)^2 - a^2\Delta]^{1/2}}{r} d\phi.$$

At the event horizon, $r = r_+$, this is

$$2\pi \frac{(r_+^2 + a^2)}{r_+} = 4\pi m,$$

which is independent of a and equal to its value in the Schwarzschild case.

Although the horizon is a surface of constant radial coordinate the proper distance around the equator is not the same as that of a circle going through the poles, which is given by

$$\int_0^{2\pi} \rho^2 d\theta.$$

Problem 3.2. *For the extreme Kerr black hole ($a = m$) show that the circumference of the event horizon in the equatorial plane is greater than the polar circumference and evaluate the latter approximately. (You will need to approximate an integral or evaluate it numerically.)*

3.2.2 *The area of the event horizon*

An important result in the classical theory of black holes is that the area of an event horizon cannot decrease, for example during a process of merger or of energy extraction. We shall defer a proof of this to the next chapter. Here we obtain an expression for the area A_h of the event horizon from the metric. The radius of the event horizon is $r_+ = m + (m^2 - a^2)^{1/2}$, and so, directly from equation (3.4), $A = (r^2 + a^2)^2 - a^2\Delta \sin^2\theta = 4m^2[m + (m^2 - a^2)^{1/2}]^2$ on the horizon. The area A_h is therefore

$$Area = A^{1/2} \int_0^{2\pi} d\phi \int_{-\pi/2}^{\pi/2} d\theta \sin\theta,$$

or,

$$A_h = 8\pi m[m + (m^2 - a^2)^{1/2}] \tag{3.9}$$

(since the angular integral is just the area of the unit sphere, or 4π).

Problem 3.3. *Putting $a = 0$ in the metric equation (3.5) we have seen that we recover the Schwarzschild metric for a mass m. But what do we get if we put $m = 0$ keeping $a \neq 0$? By making the following coordinate transformation on the Kerr metric with $m = 0$,*

$$x = \sqrt{r^2 + a^2}\sin\theta\cos\phi, \quad y = \sqrt{r^2 + a^2}\sin\theta\sin\phi, \quad z = r\cos\theta,$$

show that we recover flat spacetime.

3.3 Properties of the Kerr metric coefficients

In this section we look at some of the basic properties of the metric coefficients in Boyer–Lindquist coordinates. We shall see that the fact that the Kerr metric is not diagonal makes it somewhat more complicated to work with than the Schwarzschild metric.

3.3.1 *Identities*

By inspection of equation (3.5) we see that the metric coefficients take the following form:

$$g_{00} = \frac{(\Delta - a^2\sin^2\theta)}{\rho^2}, \tag{3.10}$$

$$g_{30} = g_{03} = \frac{2mar}{\rho^2}\sin^2\theta,$$

$$g_{11} = -\frac{\rho^2}{\Delta}, \quad g_{22} = -\rho^2, \quad g_{33} = -\frac{A\sin^2\theta}{\rho^2}.$$

The metric coefficients satisfy the following useful identities:

$$g_{33} + a\sin^2\theta\, g_{03} = -(r^2 + a^2)\sin^2\theta$$

$$g_{03} + a\sin^2\theta\, g_{00} = a\sin^2\theta$$

$$a g_{33} + (r^2 + a^2)g_{03} = -\Delta a\sin^2\theta$$

$$a g_{03} + (r^2 + a^2)g_{00} = \Delta$$

$$(g_{03})^2 - g_{00}g_{33} = \Delta\sin^2\theta.$$

All of these are proved by substitution from (3.10) and the definitions of Δ, ρ and A.

Problem 3.4. *Verify the last of these identities.*

3.3.2 *Contravariant components*

Since the Kerr metric is not diagonal, on account of the non-zero g_{03} ($= g_{30}$) term, the inverse metric cannot be found by taking the inverses of each of the components. To get the contravariant components of the metric tensor it is instead necessary to take the correct matrix inverse of $(g_{\mu\nu})$. This is most easily achieved by considering $(g_{\mu\nu})$ as a diagonal 2×2 block matrix of 2×2 matrices (with row and column order (t, ϕ, r, θ)) and taking the inverse of the diagonal blocks. The result for the non-zero entries is

$$g^{00} = \frac{(r^2 + a^2)^2 - a^2 \Delta \sin^2 \theta}{\rho^2 \Delta},$$

$$g^{03} = \frac{2mar}{\rho^2 \Delta} = g^{30},$$

$$g^{11} = -\frac{\Delta}{\rho^2},$$

$$g^{22} = -\frac{1}{\rho^2},$$

$$g^{33} = -\frac{\Delta - a^2 \sin^2 \theta}{\Delta \rho^2 \sin^2 \theta}.$$

These satisfy the useful identities

$$(g^{03})^2 - g^{00} g^{33} = \frac{1}{\Delta \sin^2 \theta},$$

$$a g^{00} - (r^2 + a^2) g^{03} = a,$$

$$(r^2 + a^2) g^{33} - a g^{03} = -\frac{1}{\sin^2 \theta}.$$

Furthermore, from

$$g_{\mu\sigma} g^{\mu\tau} = \delta^\tau_\sigma,$$

with $\sigma = 0$, $\tau = 3$ we deduce

$$\frac{g_{30}}{g_{00}} = -\frac{g^{03}}{g^{33}}.$$

Similarly, with $\sigma = 3$ and $\tau = 0$ we get

$$\frac{g^{03}}{g^{00}} = -\frac{g_{03}}{g_{33}}.$$

3.4 Interpretation of m, a and geometric units

As we have already seen, the Kerr metric (3.5) reduces to the Schwarzschild metric as $a \to 0$. To be consistent we expect that m must be the mass of the source of the gravitational field as measured by a distant orbiting body in geometric units (such that $G = 1$ and $c = 1$). The mass in conventional units would be $c^2 m / G$. (But once again, note this is not the mass of an extended material object in the system, because the Kerr solution does not correspond to the metric outside a material body.)

As $r \to +\infty$ the Kerr metric (3.5) becomes

$$d\tau^2 = \left(1 - \frac{2m}{r}\right) dt^2 + \frac{4am\sin^2\theta}{r} dt d\phi - \left(1 + \frac{2m}{r}\right) dr^2$$
$$-r^2 d\theta^2 - r^2 \sin^2\theta d\phi^2. \tag{3.11}$$

This looks like the weak field of a slowly rotating body if we put $a = j/m$. Thus, from the point of view of an orbiting body at a large distance, the parameter a is the angular momentum per unit mass in geometric units, or ca is the angular momentum per unit mass in conventional units. We shall find other strong field effects that confirm this interpretation of a.

Problem 3.5. *Derive equation (3.11) from the Kerr metric.*

3.5 Extreme Kerr black hole

We shall find that the properties of the Kerr solution differ markedly according as $a < m$, $a = m$ or $a > m$. In fact, $a > m$ is unphysical for various reasons (Section 3.18), in recognition of which the case $a = m$ is referred to as an *extreme* (or *extremal*, or *maximal*) Kerr black hole.

Problem 3.6. *By looking at the expression for the radius of the event horizon (3.8) show that a Kerr black hole with $a > m$ has no event horizon.*

3.6 Robinson's theorem

Just as the Schwarzschild metric is the unique static vacuum solution of Einstein's equations (Israel's theorem) so the Kerr metric is the unique stationary axisymmetric vacuum solution. This result is sometimes called Robinson's theorem or the Carter–Robinson theorem. Strictly both of these results require some further technical assumptions that amount to requiring the spacetime to be asymptotically flat and non-singular in the exterior of an event horizon. If the spacetime is taken to be that of a black hole, then the assumption of axisymmetry can be dropped: stationary black holes must be axisymmetric.

The uniqueness theorems can be extended to include electromagnetic fields in the environment, hence to charged black holes and (at least in theory) black holes with magnetic monopole charges, so the most general black hole is defined by four scalar quantities: mass, angular momentum and electric and magnetic charges. The fact that there are no long range multipole fields attached to an isolated black hole is sometimes expressed by saying that 'black holes have no hair'.

3.7 Particle orbits in the Kerr geometry

In the following sections we explore the spacetime of the Kerr black hole by investigating the orbits of particles and the trajectories of light rays in the exterior of the hole. Just as in the Schwarzschild case we are interested in the behaviour of geodesics, which are the paths of particles falling freely under gravity. We use a similar short-cut based on the symmetries of the metric to obtain first integrals of the motion. From these we can find the motion under a variety of initial conditions. The results are quite different from any Newtonian analogue. A rotating Newtonian body adopts a non-spherical shape as a result of its rotation. Its gravitational field differs from that of a non-rotating body only because of this different shape. A non-rotating body of the same shape would have the same Newtonian gravitational field. A rotating relativistic body influences the surrounding matter in addition directly through its rotation. Near a rotating black hole this leads to some unexpected effects.

3.7.1 *Constants of the motion*

Consider a freely falling test particle with 4-velocity u^μ in the exterior of a Kerr black hole. Just as for the Schwarzschild solution (Section 2.3.3) the covariant component of 4-velocity in a direction of symmetry is a constant. For the Kerr metric the symmetry directions in Boyer–Lindquist coordinates are $(k^\mu) = (1, 0, 0, 0)$ and $(k^\mu) = (0, 0, 0, 1)$. Thus, with E and

L_z constants (and $c = 1$)

$$u_0 = E, \tag{3.12}$$
$$u_3 = -L_z.$$

The parameter E is the energy of the particle per unit mass, and L_z is the component of angular momentum of the particle, per unit mass, normal to the equatorial plane. Note that we choose u_3 to be negative to correspond to motion in the direction of increasing ϕ. This follows because

$$u^3 = \frac{d\phi}{d\tau} = g^{30}u_0 + g^{33}u_3$$

is positive for all r if u_3 is negative (since g^{33} is negative and the other terms are each positive). Let us emphasise again that E and L_z are the energy and angular momentum per unit mass measured by a distant stationary observer. A local observer will measure a different value for the energy of a particle as we showed for the Schwarzschild case in Section 2.3.8. We should stress that L_z is the angular momentum per unit mass in the z-direction and not the total angular momentum as in the Schwarzschild case. In the Schwarzschild metric particles move in a plane and all planes are equivalent, so we were free to orient the coordinate system in order that the plane is always the equatorial plane and the angular momentum is the total. This enabled us to ignore the θ-geodesic equation. In the Kerr metric there is less symmetry and the equatorial plane is special. Thus when we interpret the constant in the limit of large r we get

$$u_3 = g_{33}u^3 = -r^2 \sin^2 \theta d\phi/d\tau = -L_z,$$

and only for orbits in the equatorial plane is L_z the total angular momentum per unit mass.

There is a further constant of the motion related to the total angular momentum that is referred to as the Carter integral, Q. It can be shown that

$$u_2 = \pm \left\{ Q - \cos^2 \theta \left[a^2(1 - E^2) + \frac{L_z^2}{\sin^2 \theta} \right] \right\}^{1/2}. \tag{3.13}$$

The Carter integral Q must be zero for motion in the equatorial plane. This follows because the axial symmetry picks out the equatorial plane as special, so a particle moving in the equatorial plane remains there, i.e. if $\theta = \pi/2$, then $d\theta/d\tau = u^2 = u_2 = 0$. There is no reason why other orbits should be confined to planes, and indeed they are not as we show in Section 3.7.4.

We shall see that orbits out of the equatorial plane are dragged round by the rotation of the hole.

Finally, the components of 4-velocity are related by

$$g^{\mu\nu} u_\mu u_\nu = 1$$

as usual (with $c = 1$).

3.7.2 *Energy*

Recall that we obtain the equations of motion of a test particle in the following way. We know that certain covariant components of the 4-velocity are constant; on the other hand, it is the contravariant components that are related to the coordinate differentials. We therefore use the metric to connect the two sets of components. For the energy equation we have

$$\frac{dt}{d\tau} = u^0 = g^{00} u_0 + g^{03} u_3.$$

By substituting for the metric coefficients from Section (3.3.2) and using (3.12), we get

$$\frac{dt}{d\tau} = \frac{AE}{\rho^2 \Delta} - \frac{2mar}{\rho^2 \Delta} L_z \tag{3.14}$$

or

$$\rho^2 \frac{dt}{d\tau} = a(L_z - aE \sin^2 \theta) + \frac{r^2 + a^2}{\Delta}[E(r^2 + a^2) - aL_z], \tag{3.15}$$

where L_z is positive for rotation of the particle in the same sense as the hole and negative in the opposite sense. This is the equation for the conservation of relativistic energy in contravariant form.

3.7.3 *Angular momentum*

For the angular momentum equation we proceed similarly. We have

$$\frac{d\phi}{d\tau} = u^3 = g^{30} u_0 + g^{33} u_3,$$

from which

$$\rho^2 \frac{d\phi}{d\tau} = \frac{L_z}{\sin^2 \theta} - aE + \frac{a}{\Delta}[E(r^2 + a^2) - L_z a], \tag{3.16}$$

where L_z is positive for co-rotation and negative for contra-rotation. This is the equation for the conservation of relativistic angular momentum in contravariant form. For motion confined to the equatorial plane this becomes

$$\Delta \frac{d\phi}{d\tau} = \frac{2ma}{r} E + L \left(1 - \frac{2m}{r} \right). \tag{3.17}$$

3.7.4 The Carter integral

The Carter integral gives the third equation of motion (replacing the condition that motion is in a plane in Schwarzschild). We have

$$u_2 = g_{22} u^2 = -\rho^2 \frac{d\theta}{d\tau},$$

which, with (3.13) gives

$$\rho^2 \frac{d\theta}{d\tau} = \pm \left\{ Q - \cos^2 \theta \left[a^2 (1 - E^2) + \frac{L_z^2}{\sin^2 \theta} \right] \right\}^{1/2}. \tag{3.18}$$

As we have seen, for $Q = 0$ there is a solution with $\theta = \pi/2$ for any E and L_z, which corresponds to motion in the equatorial plane. No other orbits can lie in a fixed plane. Orbits out of the equatorial plane are dragged round. For large r these orbits are approximately plane but have normals that precess slowly around the spin axis of the hole. (See Section 3.14.) A simple deduction from equation (3.18) is that for $E \le 1$, Q must be positive (or zero) for a solution to exist. (The square root must be real.) We shall see later that $E \le 1$ corresponds to bounded motion. These orbits move between maximum and minimum values of θ (two turning points at which $d\theta/d\tau = 0$). Also, orbits exist with $E \ge 1$. Orbits with $L_z = 0$ and $Q = (1 - E^2)a^2 \cos^2 \theta$ correspond to particles with zero total angular momentum (see Section 3.8.1). Unbound orbits with $E > 1$ can have negative Q.

We can get some insight into the physical interpretation of the Carter integral Q by considering the motion of a particle at a large distance from the black hole (Frolov and Novikov, 1998). At large r the Kerr metric simplifies to the form given in equation (3.5). Since the angular part is the same as that of flat space, the total angular momentum per unit mass is

$$L^2 = r^4 \left[\left(\frac{d\theta}{d\tau} \right)^2 + \sin^2 \theta \left(\frac{d\phi}{d\tau} \right)^2 \right]. \tag{3.19}$$

Now substitute for $d\theta/d\tau$ from equation (3.18) and for $d\phi/d\tau$ from equation (3.16). At large r these two equations become

$$r^2 \frac{d\phi}{d\tau} = \frac{L_z}{\sin^2 \theta}$$

and

$$r^2 \frac{d\theta}{d\tau} = \pm[Q - a^2 \cos^2\theta(1 - E^2) - L_z^2 \cot^2\theta]^{1/2}. \tag{3.20}$$

Substituting these expressions into (3.19) gives

$$L^2 = Q + L_z^2 + a^2 \cos^2\theta(1 - E^2).$$

As $a \to 0$ this becomes

$$L^2 = Q + L_z^2. \tag{3.21}$$

This shows that only in the limit as $a \to 0$ does Q represent the component of angular momentum orthogonal to L_z.

Problem 3.7. *From equation* (3.18) *show that for bound orbits there are two (real) turning points for* θ.

3.7.5 *The radial equation*

The radial equation is most easily obtained from the scalar product $g^{\mu\nu} u_\mu u_\nu = 1$, or

$$g^{00}(u_0)^2 + 2g^{03}u_0 u_3 + g^{11}(u_1)^2 + g^{22}(u_2)^2 + g^{33}(u_3)^2 = 1.$$

Writing the equation in this form (with the contravariant metric and covariant components of velocity) simplifies the algebra because we can use the conserved quantities (3.12) and (3.13) directly. Substituting for the metric coefficients we get, after some algebra,

$$\rho^4 \left(\frac{dr}{d\tau}\right)^2 = (E^2 - 1)r^4 + 2mr^3 - [L_z^2 + Q - a^2(E^2 - 1)]r^2$$

$$+ 2m[(L_z - aE)^2 + Q]r - a^2 Q. \tag{3.22}$$

In the equatorial plane $Q = 0$ and (3.22) can be written in the form

$$\left(\frac{dr}{d\tau}\right)^2 = E^2 - V_{eff} \tag{3.23}$$

where

$$V_{eff} = \left(1 - \frac{2m}{r}\right) + \frac{[L^2 - a^2(E^2 - 1)]}{r^2} - \frac{2m(L - aE)^2}{r^3}. \tag{3.24}$$

Note that the terms in the effective potential have the same r-dependence as the terms in the Schwarzschild effective potential, equation (2.18). The potential will therefore have the same form as that described in Section (2.3.4) and depicted in Fig. 2.2. However, the potential depends on both energy E and angular momentum L so is more complicated than that for a non-spinning black hole. We recover the non-spinning case for $a = 0$. An alternative form of the radial equation (3.22), which we shall use in Sections (3.10.3), (3.14) and (3.15.1) is

$$\rho^4 \left(\frac{dr}{d\tau}\right)^2 = [E(r^2 + a^2) - aL_z]^2 - \Delta[r^2 + (L_z - aE)^2 + Q]. \tag{3.25}$$

3.8 Frame-dragging

We now introduce a surprising, and at first sight contradictory feature of rotating black holes. Consider a particle falling from infinity with zero angular momentum, $L_z = 0$. The energy equation (3.15) simplifies to

$$\rho^2 \frac{dt}{d\tau} = \frac{[(r^2 + a^2)^2 - a^2 \Delta \sin^2 \theta]}{\Delta} E \tag{3.26}$$

and the angular momentum equation (3.16) to

$$\rho^2 \frac{d\phi}{d\tau} = \frac{2mar}{\Delta} E. \tag{3.27}$$

Despite the fact that initially the particle falls radially with no angular momentum, it acquires an angular motion during infall. The angular velocity as seen by a distant observer is

$$\omega(r, \theta) = \frac{d\phi}{dt} = \frac{d\phi/d\tau}{dt/d\tau} = \frac{2mar}{(r^2 + a^2)^2 - a^2 \Delta \sin^2 \theta}, \tag{3.28}$$

which is clearly non-zero at finite radii. Note that positive a implies positive ω, so the particle acquires an angular velocity in the direction of the spin of the hole. The angular momentum remains zero (as it must, since angular momentum is conserved). The distinction between angular momentum and angular speed arises here because of the non-zero g^{03}

term in the metric. If $g^{03} = 0$ (as in Schwarzschild) then $u^3 = g^{33}u_3$, so $L_z = u_3 = 0$ implies that $u^3 = 0$, i.e. $d\phi/d\tau = 0$; in this case zero angular momentum is equivalent to zero angular speed. However if $g^{03} \neq 0$ then $u_3 = 0$ does not imply that $u^3 = 0$ (because $u^3 = g^{30}u_0 + g^{33}u_3$) and zero angular momentum is not the same as zero angular speed.

From a physical point of view we can interpret this phenomenon as a dragging round of the local inertial frames of reference by the rotating hole. An observer in a local inertial frame must find that the angular momentum of a body subject to no non-gravitational forces is conserved. We have just seen that such a frame rotates with an angular velocity ω relative to infinity, hence is dragged round with the rotation of the hole. The frame dragging diminishes towards the poles ($\theta = 0, \pi$).

At $\Delta = 0$, and $r = r_+$, the event horizon, we get for the angular velocity

$$\omega_+ = \frac{a}{2mr_+} = \omega_H \tag{3.29}$$

irrespective of the angle θ. This is called the angular velocity of the black hole. It is the angular speed with which local inertial frames of reference are dragged round on the surface of the horizon.

Problem 3.8. *Show that equation (3.29) is valid for a particle at the horizon irrespective of its angular momentum (not just for zero angular momentum as shown in the text).*

Problem 3.9. *From the relation*

$$\omega = \frac{d\phi}{dt} = \frac{u^3}{u^0},$$

and using the final identity from Section 3.3.2 show that

$$\omega = \frac{g^{30}}{g^{00}} = -\frac{g_{03}}{g_{33}}.$$

We shall need this result later in Sections 3.9 and 3.11.

Problem 3.10. *Show that an observer at infinity attributes a speed (circumference/period) of c to a point on the equator of an extreme Kerr black hole.*

Problem 3.11. *Show that the angular speed of a slowly rotating hole* $(a \ll m)$ *is*

$$\omega_+ \approx \frac{a}{4m^2}.$$

The angular momentum is $j = ma$, *and the moment of inertia is given, as usual, by* $j = I\omega_+$. *For a slowly rotating hole show that* $I \approx mr_+^2$.

3.8.1 *Free fall with zero angular momentum*

We now consider the orbit of a particle having zero total angular momentum falling inwards along an arbitrary direction towards the black hole. Setting $L_z = 0$ in equations (3.14) and (3.16) gives

$$\frac{dt}{d\tau} = \frac{AE}{\rho^2 \Delta} \tag{3.30}$$

and

$$\frac{d\phi}{d\tau} = \frac{2marE}{\rho^2 \Delta}. \tag{3.31}$$

From (3.18) we must have

$$\frac{d\theta}{d\tau} = 0. \tag{3.32}$$

This last result follows because at large r, where the effects of spin are negligible, a zero angular momentum particle will fall radially towards the centre along a path of constant θ. As equation (3.18) has no r-dependence, this condition must hold for all values of r. So, for a particle having zero angular momentum, it follows that $Q = a^2(1 - E^2)\cos^2\theta$ and therefore for $E = 1$ that $Q = 0$, for $E < 1$ that $Q > 0$ and for $E > 1$ that $Q < 0$.

From equation (3.25), on setting $L_z = 0$ and $Q = a^2(1 - E^2)\cos^2\theta$, the radial equation becomes

$$\frac{dr}{d\tau} = \pm \frac{\left[E^2 \left(r^2 + a^2\right)^2 - \Delta\left(r^2 + a^2 E^2\right)\right]^{\frac{1}{2}}}{\rho^2}. \tag{3.33}$$

Now from (3.30) and (3.33) a distant observer sees $\frac{dr}{dt} = \frac{dr/d\tau}{dt/d\tau} \to 0$ as the particle approaches the horizon at $\Delta = 0$. From (3.30) and (3.31) $\frac{d\phi}{dt} \to \omega_+$. In other words, the particle executes an ever decreasing spiral around the horizon without ever crossing it. This is reminiscent of the behaviour of a particle approaching the event horizon of a Schwarzschild black hole. Now what happens from the point of view of the infalling particle as it approaches the horizon? From (3.33) $dr/d\tau$ remains finite up to and at the horizon, so the particle crosses the horizon in a finite lapse of its own proper time τ. Also, as the horizon is approached $d\phi/d\tau \to \infty$, so an infalling observer will see the heavens spinning overhead ever faster and infinitely fast at the horizon. These two apparently contradictory accounts can be reconciled because at the horizon an infalling clock measuring proper time τ is running infinitely more slowly than a clock at infinity measuring time t, as can be seen from equation (3.30).

For a particle released from rest at infinity with zero angular momentum away from the equatorial plane, we have shown that the angle θ is constant. However, equation (3.31) tells us that $d\phi/d\tau \neq 0$, so the particle moves inwards on the surface of a cone of constant half-angle θ. The cone closes up to a straight line for $\theta = 0$: this corresponds to the particle plunging straight down the axis of rotation.

3.8.2 *Orbits with non-zero angular momentum*

So far we have considered particles with zero angular momentum. We can find the angular velocity of a particle with non-zero angular momentum as follows. As before, for the angular speed of the particle seen from infinity we have

$$\Omega = \frac{d\phi}{dt} = \frac{u^3}{u^0} = \frac{g^{33}u_3 + g^{30}u_0}{g^{00}u_0 + g^{03}u_3}.$$

We rewrite this in terms of $\omega = g^{30}/g^{00}$:

$$\Omega = \omega - \frac{g^{30}}{g^{00}} + \frac{g^{33}u_3 + g^{30}u_0}{g^{00}u_0 + g^{03}u_3} = \omega - \frac{[(g^{30})^2 - g^{00}g^{33}]u_3}{(g^{00})^2(u_0 + \omega u_3)},$$

or, finally, using the identities in Section 3.3.2

$$\Omega = \omega - \frac{\rho^4 \Delta}{A^2 \sin^2 \theta} \frac{u_3}{(u_0 + \omega u_3)}. \tag{3.34}$$

This result applies both to massive and massless particles, with u_3 negative for co-rotation and positive for counter-rotation of the particle with respect to the hole. Note that at the horizon ($\Delta = 0$) all particles, irrespective of their angular momentum, orbit with angular velocity ω.

For the special case of a particle projected in the azimuthal direction in the equatorial plane the 4-velocity of the particle is $u_\mu = (u_0, 0, 0, u_3)$.

Expanding the scalar product $g^{\mu\nu}u_\mu u_\nu = 1$ gives

$$g^{00}(u_0)^2 + 2g^{03}u_0 u_3 + g^{33}(u_3)^2 - 1 = 0.$$

Now solve this quadratic for u_0 to get

$$u_0 = -\omega u_3 \pm \left[\left((g^{03})^2 - g^{00}g^{33}\right)u_3^2 + g^{00}\right]^{1/2}/g^{00}.$$

Using the last identity from Section 3.3.1 and $L_z = -u_3$ gives

$$u_0 + \omega u_3 = \frac{r^2 \Delta^{1/2}}{A}\left(1 + \frac{A}{r^2 L^2}\right)^{\frac{1}{2}} u_3.$$

Substituting for $u_3/(u_0 + \omega u_3)$ in equation (3.34) gives

$$\Omega = \omega \pm \frac{r^2 \Delta^{\frac{1}{2}}}{A\left(1 + \frac{A}{r^2 L^2}\right)^{\frac{1}{2}}}. \tag{3.35}$$

This is the instantaneous angular velocity of a particle projected in the azimuthal direction having angular momentum per unit mass $L_z = L$. The limiting case of a photon occurs when $L \to \infty$. We shall use this result in Section 3.10.2.

3.9 Zero angular momentum observers (ZAMOs)

In Schwarzschild spacetime we used a convenient set of local observers that 'hovered' at fixed coordinate positions. These observers have 4-velocity

$$(u^\mu) = (dt/d\tau, 0, 0, 0) = ((g_{00})^{-1/2}, 0, 0, 0)$$

and their world lines are orthogonal to surfaces of constant coordinate time t: if $(dx^\mu) = (0, dr, d\theta, d\phi)$ is a displacement in the surface $t = $ constant then $u_\mu dx^\mu = 0$.

In the Kerr spacetime the world line of a hovering observer having fixed spatial coordinates is not orthogonal to surfaces of constant Boyer–Lindquist time because of the non-zero cross term in the metric: we have

$$u_\mu dx^\mu = g_{\mu\nu}u^\nu dx^\mu = g_{03}(g_{00})^{-1/2}d\phi \neq 0.$$

Therefore, for the hovering observers, events at the same (Boyer–Lindquist) time t are not simultaneous. In order to have a convenient set of observers for which events at the same time t are simultaneous, we define a class of zero angular momentum observers (ZAMOs). These observers have angular speed ω about the spin axis, maintaining their r and θ coordinates constant.

So they are not in free fall. They correspond to the hovering observers in the Schwarzschild case. We shall first find their four velocity and then demonstrate that it is orthogonal to the surfaces of constant t.

Consider an observer in a circular orbit with zero angular momentum: $L_z = 0$, hence $u_3 = 0$. Then

$$0 = u_3 = g_{33}u^3 + g_{03}u^0,$$

or

$$u^3 = -\frac{g_{30}}{g_{33}}u^0. \tag{3.36}$$

But from the identities of Section 3.3.2 $-g_{30}/g_{33} = g^{30}/g^{00}$, which from problem 3.9 is just the angular velocity ω of an inertial frame. Thus $u^3 = \omega u^0$ for a ZAMO.

To find u^0 we use $1 = u^\mu u_\mu = u^0 u_0$ (since $u_3 = u_2 = u_1 = 0$) and

$$u_0 = g_{00}u^0 + g_{03}u^3 = (g_{00} - \frac{g_{30}}{g_{33}}g_{03})u^0.$$

This gives

$$(u^0)^2 = \frac{g_{33}}{g_{00}g_{33} - (g_{03})^2} = -\frac{g_{33}}{\Delta \sin^2 \theta}$$

and hence

$$(u^\mu)_{\text{ZAMO}} = \left(\frac{-g_{33}}{\Delta \sin^2 \theta}\right)^{1/2} (1, 0, 0, \omega). \tag{3.37}$$

We can now compute the scalar product $dx_\mu(u^\mu)_{\text{ZAMO}}$:

$$dx_\mu(u^\mu)_{\text{ZAMO}} = (g_{03}u^0 + g_{33}u^3)d\phi = 0$$

from (3.36). If follows that the world lines of zero angular momentum observers are orthogonal to the surfaces of constant t.

Problem 3.12. *For $a = 0$ show that $(u^\mu)_{\text{ZAMO}}$ is the 4-velocity of the Schwarzschild hovering observers.*

3.9.1 *Some applications of ZAMOs*

We find the local energy per unit mass relative to a ZAMO, γ_{ZAMO}, of a freely falling particle with zero angular momentum at some radius r.

The 4-velocity of a zero angular momentum particle has the form $(u_\mu) = (E, u_r, u_\theta, 0)$ and hence

$$\gamma_{\text{ZAMO}} \equiv (u^\mu)_{\text{ZAMO}} u_\mu$$

$$= E \left(\frac{g_{33}}{-\Delta \sin^2 \theta} \right)^{1/2}.$$

As we approach the horizon, $\Delta \to 0$, so $\gamma_{\text{ZAMO}} \to \infty$ i.e. the particle moves towards the horizon with a speed approaching the speed of light. This is just the result we found for the velocity of a freely falling particle with respect to a hovering observer in the Schwarzschild metric.

Problem 3.13. *Show that the velocity of a ZAMO with respect to a hovering observer in the equatorial plane is*

$$v = \frac{2ma}{r\Delta^{1/2}}$$

in the ϕ-direction. (Hint: $u^\mu_{\text{ZAMO}} u_\mu = \gamma_{\text{ZAMO}} = (1 - v^2)^{-1/2}$, and the motion is entirely in the ϕ-direction.) The speed is unity (i.e. the speed of light) at the surface $r = 2m$, which we shall see later is the static limit surface.

As a second example we shall find the energy per unit mass $E = u_0$ relative to an observer at infinity of a ZAMO for which by definition $u_3 = 0$. The time component of $u^\mu = g^{\mu\nu} u_\nu$ is

$$u^0 = g^{00} u_0 + g^{03} u_3 = g^{00} E.$$

Hence

$$E = \left(\frac{-g_{33}}{\Delta \sin^2 \theta} \right)^{\frac{1}{2}} \left(\frac{\rho^2 \Delta}{A} \right) = \left(\frac{\rho^2 \Delta}{A} \right)^{\frac{1}{2}},$$

where u^0 comes from equation (3.37).

Note that with $a = 0$ and $\theta = \pi/2$, $E = (1 - 2m/r)^{1/2}$, which is the energy of a hovering observer in the Schwarzschild metric. This futher emphasises that ZAMOs are the natural extension of the Schwarzschild hovering observer to the Kerr metric. Notice also that the energy per unit mass of a ZAMO at the horizon is zero. We shall use this result later in Section 3.15.4.

For $r_+ > r > r_-$ we have $\Delta < 0$, so the energy of a ZAMO becomes imaginary on crossing the horizon; in other words ZAMOs maintaining constant r cannot exist in this region.

Problem 3.14. *For the case $a > m$ show that ZAMOs can exist at all values of r. This shows that the Kerr solution with $a > m$ has no horizon and hence that the singularity is naked.*

3.10 Photon orbits

We now consider the motion of photons in the Kerr geometry. The equations of motion are obtained in a manner analogous to that for massive particles. From the symmetries of the metric $k^\mu p_\mu = \text{constant}$, where p is the momentum 4-vector of a photon and k is a vector pointing along a direction of symmetry. For the Kerr metric directions of symmetry are $(k^\mu) = (1, 0, 0, 0)$ and $(k^\mu) = (0, 0, 0, 1)$ so we have

$$p_0 = E_{ph}$$
$$p_3 = -L_{z,ph}$$

where E_{ph} and $L_{z,ph}$ are respectively the conserved energy and the z-component of angular momentum.

Multiplying equation (3.25) by m_0^2, the radial equation for a particle of mass m_0 has the form

$$m_0^2 \rho^4 \left(\frac{dr}{d\tau}\right)^2 = [m_0 E(r^2 + a^2) - a m_0 L_z]^2 - \Delta[m_0^2 r^2 + (L_z - aE)^2 + m_0^2 Q].$$

Taking the limit $m_0 \to 0$ and $d\tau \to 0$ with $d\lambda = d\tau/m_0$ finite, we obtain the radial equation for light rays (i.e. the equation of motion for photons)

$$\rho^4 \left(\frac{dr}{d\lambda}\right)^2 = [E_{ph}(r^2 + a^2) - a L_{z,ph}]^2 - \Delta[(L_{z,ph} - aE)^2 + Q_{ph}] \quad (3.38)$$

where $E_{ph} = m_0 E$ and $L_{z,ph} = m_0 L$ are the conserved energy and angular momentum for light. Similarly the other geodesic equations for light rays are

$$\frac{dt}{d\lambda} = \frac{A}{\rho^2 \Delta} E_{ph} - \frac{2mar}{\rho^2 \Delta} L_{z,ph} \quad (3.39)$$

$$\frac{d\phi}{d\lambda} = \frac{2mar}{\rho^2 \Delta} E_{ph} + \frac{\Delta - a^2 \sin^2 \theta}{\rho^2 \Delta \sin^2 \theta} L_{z,ph} \quad (3.40)$$

$$\rho^2 \frac{d\theta}{d\lambda} = \pm \left\{ Q_{ph} + \cos^2 \theta \left[a^2 E_{ph}^2 - \frac{L_{z,ph}^2}{\sin^2 \theta} \right] \right\}^{\frac{1}{2}}. \quad (3.41)$$

From equation (3.38) the condition of a turning point ($dr/d\lambda = 0$) is

$$E_{ph}(r^2 + a^2) - aL_{z,ph} = \Delta^{\frac{1}{2}}[(L_{z,ph} - aE)^2 + Q_{ph}]^{\frac{1}{2}}.$$

So,

$$r^2 = \Delta^{\frac{1}{2}}[(b-a)^2 + Q_{ph}/E_{ph}]^{\frac{1}{2}} + ab - a^2, \tag{3.42}$$

where $b = L_{ph}/E_{ph}$ is the impact parameter. If $\Delta < 0$ then (3.42) has no real solution for r, hence $dr/d\lambda$ cannot be zero. Therefore once a photon has moved inside $\Delta = 0$ (r decreasing, $dr/d\lambda < 0$) it cannot turn round and escape to larger r. This suggests that $\Delta = 0$ is the event horizon.

This is not an operational definition of the event horizon; in fact no local quantity can be used to determine an event horizon in general. Consider a photon that can just escape from a hole that at a later time accretes more matter. This photon, which was apparently outside the horizon, can no longer escape and was therefore, in fact, inside the event horizon. Thus the local condition, here $\Delta = 0$, defines an *apparent horizon* from which both ingoing and outgoing light rays converge to smaller radii. Only for the non-dynamical case (where the properties of the hole are fixed) does it also define the true event horizon.

3.10.1 *Photons with zero angular momentum*

From equations (3.38)–(3.41) setting $L_{z,ph} = 0$, we obtain

$$\rho^4 \left(\frac{dr}{d\lambda}\right)^2 = \left[E_{ph}(r^2 + a^2)\right]^2 - \Delta \left[a^2 E_{ph}^2 + Q_{ph}\right] \tag{3.43}$$

$$\frac{dt}{d\lambda} = \frac{A}{\rho^2 \Delta} E_{ph} \tag{3.44}$$

$$\frac{d\phi}{d\lambda} = \frac{2mar}{\rho^2 \Delta} E_{ph} \tag{3.45}$$

$$\rho^2 \frac{d\theta}{d\lambda} = \pm \left[Q_{ph} + a^2 E_{ph}^2 \cos^2 \theta\right]. \tag{3.46}$$

Employing the argument used in Section 3.8.1 for particles with zero angular momentum, we deduce that a photon will move inwards on a surface of constant θ. Therefore $Q = -a^2 E_{ph}^2 \cos^2 \theta$ and the radial equation becomes

$$\frac{dr}{d\lambda} = \pm \frac{A^{\frac{1}{2}} E_{ph}}{\rho^2}. \tag{3.47}$$

As an application, consider a spaceship plunging from rest at infinity (hence with zero angular momentum) into a Kerr black hole on a cone making an angle θ with respect to the axis of rotation. We want to find the redshift of a signal from a distant source received by the spaceship as it crosses the event horizon. Thus we need to calculate the quantity $P_\mu U^\mu$, the product of the 4-velocity of the spaceship and the 4-momentum of the photon.

We have, from equations (3.30)–(3.33), with $L_z = 0$ and $E = 1$:

$$U^\mu = \left(\frac{A}{\rho^2 \Delta}, -\frac{[2mr(r^2 + a^2)]^{1/2}}{\rho^2}, 0, \frac{d\phi}{dt} \right).$$

From equations (3.43)–(3.46) for a photon at the same conical angle θ and energy E_∞ at infinity

$$P^\mu = \left(\frac{AE_\infty}{\rho^2 \Delta}, \frac{E_\infty (r^2 + a^2)}{\rho^2}, 0, \frac{2marE_\infty}{\rho^2 \Delta} \right)$$

from which

$$P_3 = g_{33}P^3 + g_{30}P^0 = -\frac{A \sin^2 \theta}{\rho^2} \frac{2marE_\infty}{\rho^2 \Delta} + \frac{2mar \sin^2 \theta}{\rho^2} \frac{AE_\infty}{\rho^2 \Delta} = 0$$

and we obtain

$$P_\mu = \left(E_\infty, \frac{A^{1/2} E_\infty}{\Delta}, 0, 0 \right).$$

The energy in the rest frame of the spaceship is

$$E_{rec} = U^\mu P_\mu = \frac{AE_\infty}{\rho^2 \Delta} - \frac{[2mr(r^2 + a^2)]^{1/2}}{\rho^2 \Delta} \frac{A^{1/2} E_\infty}{\rho^2 \Delta}.$$

So

$$E_{rec} = \frac{E_\infty A^{1/2}}{\rho^2 \Delta}[A^{1/2} - B^{1/2}] = \frac{E_\infty A^{1/2}}{\rho^2 \Delta} \left[\frac{A - B}{A^{1/2} + B^{1/2}} \right],$$

where $B = 2mr(r^2 + a^2)$. We can simplify this since:

$$A - B = (r^2 + a^2) - a^2 \Delta \sin^2 \theta - 2mr(r^2 + a^2) = \Delta(r^2 + a^2 \cos^2 \theta) = \Delta\rho^2,$$

which gives us

$$E_{rec} = E_\infty \left[\frac{A^{1/2}}{A^{1/2} + B^{1/2}} \right].$$

At the horizon, $\Delta = 0$ so $A = B$ and $E_{rec} = E_\infty/2$. The result is interestingly independent of both the spin of the hole a and the angle θ.

3.10.2 *Azimuthal motion*

A light ray projected in the azimuthal direction in the equatorial plane has an instantaneous angular velocity given by (3.35) in the limit as $L \to \infty$

$$\Omega = \omega \pm \frac{r^2 \Delta^{1/2}}{A}. \tag{3.48}$$

Note that at the horizon $\Delta = 0$ so all particles rotate with the hole at a rate $\Omega = \omega_H$ (given by equation (3.29)), irrespective of their angular momentum.

We also see the possibility that counter-rotating photons at some radius can have zero angular velocity as seen from infinity. This means that the frame-dragging there is so strong that even photons cannot 'move backwards' (i.e. against the direction of spin of the hole). In the equatorial plane this occurs at a radius given by $A\omega = r^2\Delta^{1/2}$ or

$$A^2\omega^2 = 4m^2a^2r^2 = r^4\Delta = r^4(r^2 + a^2 - 2mr),$$

which is satisfied by $r = 2m$ (and hence $\Delta = a^2$). The surface in which counter-rotating photons appear stationary from infinity is called the static limit surface. (See Section 3.11.) We have therefore shown that the intersection of the static limit surface with the equatorial plane occurs at $r = 2m$.

3.10.3 *Photon capture cross-section*

Now we can look at the effect of rotation on the capture of photons by a black hole. This can done by considering circular photon orbits, which are unstable and lie on the borderline between capture by the hole and escape to infinity. For simplicity we confine our attention to motion in the equatorial plane. As in Section 2.3.9 the conditions for circular orbits are obtained from the radial equation of motion. From equation (3.25) for photons moving in the equatorial plane we get as

$$\left(\frac{dr}{d\lambda}\right)^2 = r^{-4}\left\{\left[E_{ph}(r^2 + a^2) - aL_{ph}\right]^2 - \Delta(L_{ph} - aE_{ph})^2\right\}$$

$$= E_{ph}^2 + \frac{2m}{r^3}(L_{ph} - aE_{ph})^2 - \frac{1}{r^2}(L_{ph}^2 - a^2E_{ph}^2)$$

$$= R(r).$$

The conditions for a circular orbit are, as before (Section 2.3.9), $dr/d\lambda = 0$ and $dR/dr = 0$. Applying these conditions for co-rotating photons gives

$$E_{ph}^2 + \frac{2m}{r_{ph}^3}(L_{ph} - aE_{ph})^2 - \frac{1}{r_{ph}^2}(L_{ph}^2 - a^2E_{ph}^2) = 0, \qquad (3.49)$$

and

$$-\frac{6m}{r_{ph}^4}(L_{ph} - aE_{ph})^2 + \frac{2}{r_{ph}^3}(L_{ph}^2 - a^2E_{ph}^2) = 0.$$

We can deal with counter-rotating photons by changing the sign of a. The second of these two equations gives directly

$$r_{ph} = 3m\frac{b - a}{b + a}, \qquad (3.50)$$

where $b = L/E$ is the impact parameter below which capture will occur. For $a = 0$ we regain the Schwarzschild result $r_{ph} = 3m$; for a co-rotating hole we have $a > 0$ and $r_{ph} < 3m$, with the opposite inequalities for a counter-rotating hole.

Solving for the impact parameter in terms of the radius of the circular photon orbit

$$b = -a\frac{r_{ph} + 3m}{r_{ph} - 3m},$$

so the extreme values are $b = 2m$ (for $a = m$, $r_{ph} = m$ – see below, Section 3.13.6) and $b = 7m$ (for $a = -m$, $r_{ph} = 4m$). These straddle the Schwarzschild value $b = 3\sqrt{3}m$. Thus, contra-rotating photons are more likely to be captured than co-rotating photons. We shall refer to this result in Section 3.16.1 below.

The shape of the shadow of a black hole is determined by the impact parameter for entry into a circular photon orbit (Section 2.4.2). In the present case, this parameter depends on the side of the black hole that the photons pass, and on the viewing angle. Thus, when viewed in the equatorial plane the shadow will be elongated on the side on which photons are moving against the rotation of the hole, and compressed on the opposite side. When viewed along the rotation axis the shadow is circular and at intermediate angles the degree of elongation depends on the viewing angle.

The capture cross-section for photons incident perpendicular to the spin axis is $\sigma_{perp} \approx 24.3\pi m^2$ and for photons moving in a direction parallel to the spin axis is $\sigma_{||} \approx 23.3\pi m^2$. (See Frolov and Novikov, 1998, p78, for further details.) Thus the capture cross-section for a rotating black hole is less than $\sigma = 27\pi m^2$ for a non-rotating one (Section 2.7.1).

Problem 3.15. *For circular photon orbits in the equatorial plane, show that (i) $(b+a)^3 = 27m^2(b-a)$ and (ii) $b = 3(mr)^{1/3} - a$. Hence show that*

$$\left(\frac{r_{ph}}{m}\right)^{\frac{3}{2}} - 3\left(\frac{r_{ph}}{m}\right) \pm \frac{2a}{m} = 0.$$

3.11 The static limit surface

We shall show that the static limit surface is defined by $g_{00} =, 0$, from which we shall be able to find this surface in general (not just in the equatorial plane). Consider a stationary particle, $r = $ constant, $\theta = $ constant, $\phi = $ constant. We have

$$1 = g_{00}\left(\frac{dt}{d\tau}\right)^2.$$

For $g_{00} \leq 0$ this condition cannot be satisfied, so a massive particle cannot be stationary within the surface $g_{00} = 0$. For photons we can satisfy the corresponding condition if $g_{00} = 0$, which is in agreement with what we found above, that photons can be stationary at the static limit. Solving the condition $g_{00} = 0$ for r gives us the radius of the static limit surface:

$$r_{st} = m + (m^2 - a^2\cos^2\theta)^{1/2},$$

where we take the positive square root to obtain the outer surface, and also for agreement with $r_{st} = 2m$ in the equatorial plane.

Note that the static limit surface lies outside the horizon $\Delta = 0$, except at the poles where the two surfaces coincide (see Fig. 3.1). We show next that the static limit surface in the Kerr solution is not a horizon. This is in contrast to the Schwarzschild geometry, where the event horizon and

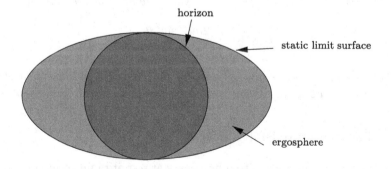

Fig. 3.1   Ergosphere of a Kerr black hole.

the static limit surface coincide. Consider a particle in a circular orbit (not necessarily a geodesic) on $\theta = $ constant. We have

$$1 = g_{00}(u^0)^2 + 2g_{03}u^0u^3 + g_{33}(u^3)^2$$
$$= (u^0)^2(g_{00} + 2\Omega g_{03} + \Omega^2 g_{33}),$$

where as usual $\Omega = d\phi/dt = u^3/u^0$. Thus, for the motion to be possible, that is, for the orbit to be timelike, we require $g_{00} + 2\Omega g_{03} + \Omega^2 g_{33} > 0$. As an example, consider a zero angular momentum particle in the equatorial plane; the inequality to be satisfied becomes

$$2\omega g_{03} + \omega^2 g_{33} + g_{00} > 0.$$

But $\omega = -g_{03}/g_{33}$ (problem 3.9) so the inequality simplifies to

$$\frac{-g_{03}^2 + g_{00}g_{03}}{g_{33}} = -\frac{\Delta}{g_{33}} > 0$$

where we have used an identity from Section 3.3.1 with $\sin\theta = 1$ and where the final inequality follows since $\Delta > 0$ and $g_{33} < 0$ outside the horizon. Thus inside the static limit there are possible circular orbits. We can perturb a circular orbit to give the particle a small velocity in the outward radial direction. Thus a particle can escape from within the static limit surface, so it is not an event horizon.

Problem 3.16. *Show that the purely radial motion of light inside the static limit is not possible.*

Problem 3.17. *Show that a stationary particle (i.e. a particle with fixed coordinates) in the Kerr metric has angular momentum per unit mass*

$$L_z = -\frac{2mar\sin^2\theta}{\rho(\Delta - a^2\sin^2\theta)^{1/2}}$$

in the opposite sense to the rotation of the hole. Hence show that it is impossible for a body to remain stationary inside the static limit surface.

3.12 The infinite redshift surface

The redshift of radiation emitted by a body at rest at radius r and observed at infinity can be obtained as follows. The metric relates the local time τ of the stationary frame at r to the time t for a stationary observer at infinity. The redshift is given in terms of the wavelengths of the radiation by

$$1 + z = \frac{\lambda_\infty}{\lambda_r}$$

so, since the wavelength is proportional to the period of the wave, the redshift also relates the two times:

$$1 + z = \frac{dt}{d\tau} = (g_{00})^{-1/2}.$$

Thus the static limit surface in the Kerr metric, $g_{00} = 0$, is also an infinite redshift surface, just as in Schwarzschild. The coincidence of these two surfaces is peculiar to these geometries and is not true for general black holes (for example, in the presence of external matter).

3.13 Circular orbits in the equatorial plane

We now consider circular orbits in the equatorial plane. As for the Schwarzschild case we shall find that there are stable bound orbits down to a limiting value of radius r inside which the circular orbits are unstable. At still smaller radii the unstable orbits become unbound and culminate in a circular photon orbit. The location of the transition between these different types of orbits depends on the angular momentum of the black hole a and the sense of rotation of the orbit (prograde or retrograde) with respect to the hole.

The radial motion of a particle is governed by (3.25)

$$\left(r^2 \frac{dr}{d\tau} \right)^2 = R(E, L, r) \tag{3.51}$$

with $Q = 0$ and $L = L_z$. For a circular orbit we require that the radius is constant, hence that $dr/d\tau = 0$, and also that $d^2r/d\tau^2 = 0$. Thus, for a circular orbit as we saw in Section 2.3.9, $R = 0$ and $dR/dr = 0$. From equation (3.25) the condition $R = 0$ gives

$$R = E^2 - 1 + \frac{2m}{r} - \frac{[L^2 - a^2(E^2 - 1)]}{r^2} + \frac{2m}{r^3}(L - aE)^2 = 0 \tag{3.52}$$

and

$$\frac{1}{2}\frac{dR}{dr} = -\frac{m}{r^2} + \frac{[L^2 - a^2(E^2 - 1)]}{r^3} + \frac{3m}{r^4}(L - aE)^2 = 0. \tag{3.53}$$

Rearranging (3.53) gives

$$mr^2 - [L^2 - a^2(E^2 - 1)]r + 3m(L - aE)^2 = 0.$$

Solving this quadratic in r, we find the radii of the circular orbits:

$$r = \frac{[L^2 - a^2(E^2 - 1)] \pm \{[L^2 - a^2(E^2 - 1)]^2 - 12m^2(L - aE)^2\}^{\frac{1}{2}}}{2m}$$

By appealing to the Schwarzschild limit ($a = 0$) we deduce that the positive sign gives the radii of the stable orbits and the negative sign the radii of the unstable orbits.

We can solve equations (3.52) and (3.53) for E and L as functions of r (Lynden-Bell, 1978). The first step is to eliminate the term that mixes L and E by multiplying (3.52) by $2r/3$ and adding this to (3.53). This gives

$$R + r\frac{dR}{dr} = (E^2 - 1)(3r^2 + a^2) + 4mr - L^2 = 0, \tag{3.54}$$

which is an explicit relation between E and L. Eliminating L from (3.53) using (3.54) gives a quadratic for $E^2 - 1$. After some considerable algebra we obtain

$$E = \frac{1 - \frac{2m}{r} \pm \frac{a}{m}\left(\frac{m}{r}\right)^{3/2}}{\left[1 - \frac{3m}{r} \pm 2\frac{a}{m}\left(\frac{m}{r}\right)^{3/2}\right]^{1/2}}. \tag{3.55}$$

Substituting (3.55) into (3.54) gives

$$L = \pm\frac{m}{\left(\frac{m}{r}\right)^{1/2}}\frac{1 + \left(\frac{a}{m}\right)^2\left(\frac{m}{r}\right)^2 \mp 2\frac{a}{m}\left(\frac{m}{r}\right)^{3/2}}{\left[1 - \frac{3m}{r} \pm 2\frac{a}{m}\left(\frac{m}{r}\right)^{3/2}\right]^{1/2}}. \tag{3.56}$$

The upper signs correspond to prograde orbits (in the same direction as the spin of the hole) and the lower signs to retrograde orbits.

3.13.1 *Innermost (marginally) stable circular orbit*

The stability of a circular orbit can be shown to depend on the sign of the second derivative of the effective potential, as one might expect. To simplify the algebra however, it is easier to proceed directly as follows. Writing L

for L_z, the radial equation in the equatorial plane again has the form (3.51):

$$r^4 \left(\frac{dr}{d\tau}\right)^2 = R(r, L, E),$$

where, from (3.25) with $Q = 0$, the function R is

$$R = \{[E(r^2 + a^2) - aL]^2 - \Delta[r^2 + (L - aE)^2]\}.$$

We can derive the condition for stability in the standard way. Assume that the particle is in the orbit $r = r_0$ and make a small perturbation to the orbit $r = r_0 + \varepsilon$. Then

$$r_0^4 \left(\frac{d\varepsilon}{d\tau}\right)^2 = R(r_0) + \varepsilon R'(r_0) + \frac{1}{2}\varepsilon^2 R''(r_0) + \cdots,$$

where the prime denotes differentiation with respect to r. The orbit $r = r_0$ is a solution of $R(r_0) = 0$ and $R'(r_0) = 0$, so ε grows exponentially

$$d\varepsilon/d\tau \simeq \left[R''(r_0)/2r_0^4\right]^{1/2} \varepsilon$$

$$\propto +\varepsilon$$

unless $R''(r_0) \leq 0$. For the marginally stable case we have $R''(r_0) = 0$. This allows us to determine the energy, angular momentum and radius of the innermost stable circular orbit (from the three conditions on R).

To simplify the algebra the trick is to note that equation (3.54) has the form $R + rR' = 0$. Thus we can differentiate again to obtain

$$2R' + rR'' = 6r(E^2 - 1) + 4m \qquad (3.57)$$

an expression that must vanish at the radius $r_0 = r_{ms}$ of a marginally stable orbit, where both $R' = 0$ and $R'' = 0$. The resulting equation (3.57) is readily solved for E^2:

$$E^2 = 1 - \frac{2}{3}\frac{m}{r_{ms}}. \qquad (3.58)$$

But we also have the condition (3.55) for $E(r)$ on a circular orbit. Eliminating E from (3.55) using (3.58) gives

$$r_{ms}^2 - 6mr_{ms} \pm 8am^{1/2}r_{ms}^{1/2} - 3a^2 = 0. \qquad (3.59)$$

This is a quartic in $r_{ms}^{1/2}$, which can be solved (Bardeen *et al.*, 1972), although the resulting algebraic expression is not very illuminating. Instead consider two special cases. First, we can confirm that we recover the Schwarzschild result $r_{ms} = 6m$ on putting $a = 0$. Second, for the extreme

Kerr solution with $a = m$, the quartic for the innermost (marginally) stable orbit at $r = r_{ms}$ reduces to

$$\left(\frac{r_{ms}}{m}\right)^2 - 6\left(\frac{r_{ms}}{m}\right) \pm 8\left(\frac{r_{ms}}{m}\right)^{1/2} - 3 = 0.$$

With the upper sign this is satisfied by $r_{ms} = m$; this is the marginally stable orbit for a co-rotating particle. With the lower sign we see that $r_{ms} = 9m$, which is the solution for a counter-rotating particle.

Problem 3.18. *It is difficult to solve equation (3.59) for r_{ms} but it can be solved readily as an equation for a, the angular momentum of the hole, as a function of r_{ms}. Hence verify that for retrograde orbits $r_{ms} = 7.5m$ corresponds to $a = 0.48m$ and for prograde orbits $r = 4.12m$ corresponds to $a = 0.53m$.*

3.13.2 *Period of a circular orbit*

The coordinate period of a particle in orbit in the equatorial plane of the Kerr metric is important astrophysically because this is a timescale on which periodic variability in the emission of radiation from material in orbit about the hole, for example in an accretion disc, would be detected by a distant observer. We can calculate this period as follows. We have

$$\frac{d\phi}{dt} = \frac{d\phi}{d\tau}\frac{d\tau}{dt},$$

where the quantities on the right are independent of time t and angle ϕ. Thus, integrating over a period T, $0 \leq \phi \leq 2\pi$,

$$2\pi = \left(\frac{d\phi}{d\tau}\frac{d\tau}{dt}\right)T$$

or

$$T = 2\pi\frac{dt}{d\tau}\Big/\frac{d\phi}{d\tau}.$$

Using equations (3.14) and (3.17) with E and L given by (3.55) and (3.56) we get, after some algebra,

$$T = 2\pi m \left(\frac{r}{m}\right)^{3/2}$$

$$\times \frac{\left[1 + \frac{a^2}{r^2} - \frac{2m}{r} \pm \left(\frac{a}{m}\right)\left(\frac{m}{r}\right)^{3/2} \pm \left(\frac{a}{m}\right)^3 \left(\frac{m}{r}\right)^{7/2} \mp 2\left(\frac{a}{m}\right)\left(\frac{m}{r}\right)^{5/2}\right]}{\left[1 + \frac{a^2}{r^2} - \frac{2m}{r}\right]}.$$

which, on dividing the numerator by the denominator, simplifies to

$$T = 2\pi \left(\frac{r^{\frac{3}{2}}}{m^{\frac{1}{2}}} \pm a \right). \tag{3.60}$$

In the Schwarzschild case, $a = 0$, we get

$$T = 2\pi m \left(\frac{r}{m} \right)^{3/2}$$

and the last stable orbit is at $r_{ms} = 6m$, so the minimum period is $T \approx 29.4\pi m$. For the black hole at the centre of the Galaxy, with a mass $4.0 \times 10^6 M_\odot$, this period is about 30 minutes. At the opposite extreme, if the black hole at the Galactic centre were an extreme Kerr hole with material in prograde orbits then $a = m$, $r_{ms} = m$ and $T = 4\pi m$, giving a period of 3.8 minutes. For extreme retrograde orbits, then $a = m$, $r_{ms} = 9m$ and $T = 52\pi m$, giving a period of about 48 minutes. Note that most of the difference between these periods arises from the change to r_{ms} from m to $9m$ rather than the explicit dependence of T on a in equation (3.60).

It is interesting to compare the proper periods of two clocks orbiting in opposite senses in the equatorial plane of a spinning black hole at radius r. From equations (3.14) we have

$$\frac{dt}{d\tau} = \frac{AE}{r^2 \Delta} - \frac{2mar}{r^2 \Delta} L,$$

where τ is the proper time of the orbiting clock. Substituting for E and L from equations (3.55) and (3.56) we get, after some algebra, the following relation between the time $\Delta\tau$ recorded by an orbiting clock at a radius r and the time Δt recorded by a distant stationary clock:

$$\Delta\tau = \frac{\left[1 - \frac{3m}{r} \pm \frac{2a}{m} \left(\frac{m}{r} \right)^{\frac{3}{2}} \right]^{\frac{1}{2}}}{\left[1 \pm \frac{a}{m} \left(\frac{m}{r} \right)^{\frac{3}{2}} \right]} \Delta t,$$

where, as usual, the upper sign refers to the prograde motion and the lower sign to retrograde motion. It is straightforward to show that, measured by a distant clock, $\Delta\tau_+ > \Delta\tau_-$ (see problem 3.20). The result can also be seen by considering an extreme Kerr black hole in the particular case $r \to 4m$. In this limit the retro-rotating clock becomes a photon with $\tau_- \to 0$ whilst τ_+ remains finite.

This is an example of the 'twin paradox' but in contrast to the effect in special relativity both twins here follow geodesic paths.

Problem 3.19. *Find the period of a particle in the last stable prograde orbit about the black hole at the centre of the Galaxy assuming a = 0.53 and r = 4.12.*

Problem 3.20. *Show that* $\Delta\tau_+ > \Delta\tau_-$ *i.e. that*

$$
\frac{\left[1 - \frac{3m}{r} + \frac{2a}{m}\left(\frac{m}{r}\right)^{\frac{3}{2}}\right]^{\frac{1}{2}}}{\left[1 + \frac{a}{m}\left(\frac{m}{r}\right)^{\frac{3}{2}}\right]} > \frac{\left[1 - \frac{3m}{r} - \frac{2a}{m}\left(\frac{m}{r}\right)^{\frac{3}{2}}\right]^{\frac{1}{2}}}{\left[1 - \frac{a}{m}\left(\frac{m}{r}\right)^{\frac{3}{2}}\right]}.
$$

3.13.3 *Energy of the innermost stable orbit*

Given a value for a we can solve the quartic equation (3.59) to get the radius r_{ms} of the innermost stable orbit. Substituting the value of r_{ms} into equation (3.58) gives the energy per unit mass of a particle in this orbit. For the limiting case of co-rotation around an extreme Kerr black hole we have $r_{ms} = m$ and thus,

$$
E = \left(1 - \frac{2}{3}\right)^{1/2} = \frac{1}{\sqrt{3}}.
$$

The binding energy of a particle in an orbit is the difference between the orbital energy and the energy of the particle at infinity. So the binding energy per unit mass is $1 - E$, or about 42% of the energy of the particle at rest at infinity. The corresponding results for a counter-rotating particle are $r_{ms} = 9m$ for the radius of the innermost stable orbit and $E = (25/27)^{1/2}$, and a binding energy of 3.8%. For comparison, recall that for a non-rotating black hole the corresponding binding energy is 5.7% of the rest mass energy.

In fact we shall see in Sections (3.16.1) and (4.4.2) that real black holes cannot be spun up to the extreme Kerr limit. A more realistic upper limit to a is given by $a = 0.998m$. Inserting this value into equation (3.59) gives $r_{ms} = 1.237m$, which yields a maximum binding energy for a co-rotating particle of about 32%. This is the maximum energy that can be realistically extracted from matter in an accretion disc as it spirals into a black hole.

Problem 3.21. *Verify, using equation (3.59), that* $r_{ms} = 1.237m$ *for* $a = 0.998m$. *Hence, using Fig. 3.2, sketch the relationship between binding energy in the innermost stable orbit and a.*

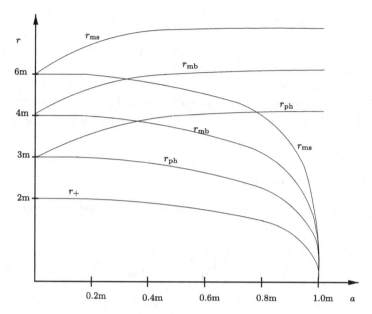

Fig. 3.2 Radii of circular orbits in the equatorial plane as a function of a. The upper curves refer to retrograde and the lower to prograde orbits. (From Bardeen *et al.*, 1972.)

3.13.4 *Angular momentum of the innermost stable orbit*

Substituting for E^2 from (3.58) in (3.54) we can obtain the angular momentum on the innermost (marginally) stable circular orbit. Using (3.59) we can eliminate a to get

$$L = \pm \frac{2m}{3\sqrt{3}} \left\{ 1 + 2 \left[3 \left(\frac{r_{ms}}{m} \right) - 2 \right]^{1/2} \right\}. \tag{3.61}$$

Problem 3.22. *Derive equation (3.61).*

Problem 3.23. *Find the angular momentum carried into an extreme Kerr black hole from the marginally stable orbit by (a) a co-rotating particle and (b) a counter-rotating particle.*

3.13.5 *Marginally bound orbits*

Bound orbits correspond to $E \leq 1$. We can calculate the radius of a marginally bound orbit by putting $E = 1$ in (3.55). Solving the resulting quartic equation gives

$$r_{mb} = 2m \mp a + 2m^{1/2}(m \mp a)^{1/2} \tag{3.62}$$

with the upper signs corresponding to co-rotation (Bardeen *et al.*, 1972). As an example, in the case of an extreme Kerr black hole, $a = m$, for a co-rotating particle $r_{mb} = m$ and for a counter-rotating particle $r_{mb} = (3 + 2\sqrt{2})m$.

3.13.6 *Unbound orbits*

Circular orbits exist where the square root in the denominator of equation (3.55) is real, hence where

$$\left(\frac{r}{m}\right)^{3/2} - 3\left(\frac{r}{m}\right)^{1/2} \pm 2\left(\frac{a}{m}\right) \geq 0. \tag{3.63}$$

Solving this cubic in the limiting case of equality gives

$$r = r_{ph} = 2m\left\{1 + \cos\left[\frac{2}{3}\cos^{-1}\left(\mp\frac{a}{m}\right)\right]\right\}. \tag{3.64}$$

In this case the energy per unit mass $E \to \infty$, so the orbit must (if anything) correspond to a zero mass particle i.e. a photon orbit. For $a = 0$ we recover the Schwarschild result, $r_{ph} = 3m$, for the innermost photon orbit. For the extreme Kerr solution we get $r_{ph} = m$, or $4m$ (co-rotation or counter-rotation respectively). Just as in the Schwarzschild case we see that there is a region in which circular orbits can be unbound!

Problem 3.24. *For what value of a does the prograde photon orbit in the equatorial plane lie at the static limit surface?*

3.13.7 *Proper distances in the extreme Kerr black hole*

Although in the extreme Kerr solution $r_{ph} = r_{mb} = r_{ms} = m$ these orbits are distinct and separated from each other by non-zero proper distances. The explanation lies with the failure of the Boyer–Lindquist r coordinate to represent the geometry in this region, i.e. the coordinate system becomes singular at this radius (Bardeen *et al.*, 1972). See Figs. 3.2 and 3.3.

Fig. 3.3 Embedding diagram for the Kerr solution. As $a/m \to 1$, the 'throat' approaches a cylinder: the radii of the various labelled surfaces (shown as curves in this representation) become equal in magnitude, but the surfaces are nevertheless distinct.

We can demonstrate this as follows. The proper distance between the horizon and radial coordinate R is

$$D = \int_{r_+}^{R} \frac{r}{\Delta} dr,$$

where R stands for one of r_{ph}, r_{mb} or r_{ms}. We evaluate these integrals for $a = m(1 - \delta)$, where $\delta \ll 1$ and then let $\delta \to 0$ to obtain the proper distances in the extreme Kerr case. For non-zero δ

$$\Delta^{\frac{1}{2}} \sim m \left[\left(\frac{r}{m} - 1 \right)^2 - 2\delta \right]^{\frac{1}{2}},$$

so

$$D(R - r_+) = \int_{r_+}^{R} \frac{r/m}{[(r/m - 1)^2 - 2\delta]^{1/2}} dr. \tag{3.65}$$

We find r_{ph} from equation (3.63). Let $x = r/m$ and put $a = m(1 - \delta)$; then (3.63) becomes

$$x^{\frac{3}{2}} - 3x^{\frac{1}{2}} + 2(1 - \delta) = 0.$$

We now let $x = (1 + \epsilon)$ where ϵ is small, and expand by the binomial theorem. We keep terms to the second order in ϵ because the first order terms cancel. This gives

$$\frac{6}{8}\epsilon^2 - 2\delta = 0$$

or

$$r_{ph} \sim m\left[1 + \left(\frac{8\delta}{3}\right)^{1/2}\right].$$

Similarly from equation (3.62) we get

$$r_{mb} \sim m(1 + 2\delta^{1/2}).$$

The radius of marginally stability, r_{ms}, is a bit more tricky. Following the same procedure, we get

$$-2\delta - 4\delta\epsilon - 3\delta^2 + \frac{1}{2}\epsilon^3 = 0. \tag{3.66}$$

This is consistent only if

$$-2\delta + \frac{1}{2}\epsilon^3 = 0,$$

in which case the remaining terms in (3.66) are of order ϵ^4 and can be neglected. So finally

$$r_{ms} \sim \left[1 + (4\delta)^{1/3}\right].$$

Evaluating the integral in (3.65) and letting $\delta \to 0$ in each case gives

$$D(r_{ph} - r_+) = \frac{m}{2} \log 3$$
$$D(r_{mb} - r_+) = m \log(1 + \sqrt{2})$$
$$D(r_{ms} - r_+) \to \infty.$$

Note that the infinite proper distance for r_{ms} does not have a physical significance as it is a spacelike interval. This means it does not connect events on a particle worldline: an infalling particle reaches the event horizon in a finite proper time. To illustrate this consider the case of a particle with $E = 1$ and $L = 0$ falling radial from a point with coordinate R to the horizon at $r = r_+$. From equation (3.22) we obtain

$$\left(\frac{dr}{d\tau}\right)^2 = \frac{2m}{r} + \frac{2ma^2}{r^3}.$$

Setting $a = m$, separating the variables and integrating gives

$$\tau = \int_{r_+}^{R} \frac{r^{3/2}}{(2m)^{1/2}(r^2 + m^2)^{1/2}} dr.$$

If the upper limit is finite then the integral is a finite quantity integrated over a finite range and hence finite. So a particle falling from a finite radial distance reaches the horizon in a finite lapse of its own proper time.

Problem 3.25. *Evaluate the integral in equation (3.65) in the cases* $R = r_{ph}$ *and* $R = r_{mb}$.

3.14 Polar orbits

We get another view of the effects of frame dragging by a rotating black hole if we look at the precession of non-equatorial orbits. To state this precisely, we define the line of the ascending node as the line from the origin in the equatorial plane through the point at which the orbit crosses the plane in going from negative to positive latitudes. For orbits of decreasing radii we find that the line of the ascending node is increasingly dragged in the sense of the spin of the black hole. To see how this comes about, consider the simple case of an orbit of constant r that passes through the poles. Since the orbit lies on a sphere, we shall refer to it as a spherical orbit. The angle along the orbit is then θ and the precession is measured by ϕ.

For an orbit through the poles $L_z = 0$, so the radial equation, equation (3.25) becomes

$$\rho^4 \left(\frac{dr}{d\tau}\right)^2 = E^2(r^2 + a^2)^2 - \Delta[r^2 + a^2 E^2 + Q] \equiv V_r.$$

The condition for a spherical orbit is, as usual, $V_r = 0$ and

$$\frac{dV_r}{dr} = 4r(r^2 + a^2)E^2 - 2r\Delta - 2(r - m)[r^2 + a^2 E^2 + Q] = 0.$$

Eliminating the expression in square brackets between these two equations gives

$$E^2 = \frac{r\Delta^2}{(r^2 + a^2)(r^3 - 3mr^2 + a^2r + a^2m)} \tag{3.67}$$

for the energy of a particle in a polar orbit with constant r. We should note that the distinction between co- and contra-rotating particles disappears for polar orbits.

We can find the location of the marginally stable polar orbit, r_{ms}, for the limiting case of an extreme Kerr black hole, by recalling that the marginally stable orbit lies at the minimum of E. Differentiating equation (3.67) with respect to r and setting the result to zero for the minimum gives the quartic equation

$$\left(\frac{r_{ms}}{m}\right)^4 - 4\left(\frac{r_{ms}}{m}\right)^3 - 6\left(\frac{r_{ms}}{m}\right)^2 - 4\frac{r_{ms}}{m} + 1 = 0,$$

which has the physically relevant exact solution

$$r_{ms} = m\left[(1 + \sqrt{3}) + (3 + 2\sqrt{3})^{1/2}\right] \simeq 5.275m.$$

Inserting this value of r into equation (3.67) gives $E = 0.937$ so the binding energy $1 - E = 0.627$ or 6.37% of the rest energy of the particle. The binding energy of a particle in the corresponding orbit of a non-rotating black hole is 5.72% (Section 2.3.12). Evidently the binding energy of a particle in a polar orbit is only weakly dependent on the spin of the black hole. By contrast the binding energy of a particle in the equatorial plane ranges from 3.8% to about 42% depending on the magnitude and sense of the spin of the black hole (Section 3.13.3).

Only orbits having $E^2 > 0$ can exist. The numerator of (3.67) is positive for all values of r down to the horizon but the denominator will fall to zero for some value of r before the horizon is reached at which point $E^2 \to \infty$: this is the circular photon orbit.

For the limiting case of an extreme Kerr black hole we set $a = m$ in equation (3.67) and factorise the denominator to get

$$E^2 = \frac{r(r - m)^4}{(r^2 + m^2)(r^2 - 2mr - m^2)(r - m)}.$$

After cancelling the factor $(r - m)$, we see that $E^2 \to \infty$ when $r^2 - 2mr - m^2 = 0$. Solving this gives

$$r_{ph} = (1 + \sqrt{2})m.$$

Problem 3.26. *For large* r $(\gg m$ *and* $\gg a)$ *show that we recover the Schwarzschild result*

$$E^2 = \frac{(1 - 2m/r)^2}{(1 - 3m/r)}.$$ (3.68)

See Section 2.3.10.

It is interesting to note that the coordinates r_+, r_{ph} and r_{ms} are distinct for polar orbits around an extreme Kerr black hole, in contrast to the equatorial orbits (Section 3.13).

Problem 3.27. *From equation* (3.28), *in the approximation of large* r, *show that* ω_p *is given by*

$$\omega_p = \frac{2ma}{r^3}.$$

This equation is given in geometrical units so the time τ *is in metres, and* ω_p *is in* m^{-1}. *Show that* ω_p *is given in SI units by*

$$\omega_p(in\ seconds^{-1}) = \frac{2G}{c^2}\frac{J}{r^3},$$

where J *is the angular momentum of the black hole* $(= aMc$ *for a body of mass* M). *Show that for a given* r/m *(i.e. for a given distance from the hole measured in terms of Schwarzschild radii) the precession is stronger for lower mass black holes.*

Problem 3.28. *A satellite is in a circular polar orbit of radius* 12 000 *km around the Earth. Show that the plane of the orbit will be dragged round by the rotation of the Earth at a rate of about* 3×10^{-2} *arcsec per year.* (*The moment of inertia of the Earth is* $0.33 M_\oplus R_\oplus^2$.)

Problem 3.29. *Show that at large distances from a slowly rotating body a particle in the equatorial plane having zero angular momentum is dragged round at the same angular frequency as that at which a polar orbit of the same radius precesses.*

Problem 3.30. *Using equation* (3.67) *and the condition* $V_r = 0$ *show that for a circular polar orbit*

$$Q = \frac{mr^2(r^4 + 2a^2r^2 - 4ma^2r + a^4)}{(r^2 + a^2)(r^3 - 3mr^2 + a^2r + a^2m)}$$

and hence, for $a = 0$ (*the Schwarzschild case, valid for small* a *or large* r)

$$Q = \frac{mr}{(1 - 3m/r)} = L^2.$$

3.14.1 *Orbital period of polar orbits and the motion of nodes*

Whilst a particle is moving through 2π radians in θ, the ascending node of the orbit is being dragged in the sense of rotation of the black hole. The rate of change of ϕ is given by

$$\frac{d\phi}{dt} = \frac{2mar}{(r^2 + a^2)^2 - a^2\Delta\sin^2\theta}. \tag{3.69}$$

For large r, $d\phi/dt \propto 1/r^3$, so the movement in ϕ is small and the polar orbits are almost confined to a plane. However, for small r, and in the extreme case $a = m$, the dragging of the orbit is large.

To illustrate this we shall look at the limiting case of the polar photon orbit around an extreme Kerr black hole for which $r_{ph} = (1 + \sqrt{2})m$ (Section 3.14). Substituting into (3.69) gives

$$\frac{d\phi}{dt} = \frac{2r_{ph}/m}{m\left\{[(r_{ph}/m)^2 + 1]^2 - [(r_{ph}/m)^2 - 1]^2\sin^2\theta\right\}}$$

$$= \frac{4.828}{m(46.627 - 2\sin^2\theta)}. \tag{3.70}$$

The rate of change of ϕ is a function of θ: it changes over the course of an orbit from a maximum of 0.1082 rad s^{-1} at $\theta = \pi/2$ to a minimum of 0.1035 rad s^{-1} at $\theta = 0$. So an average rate of change is

$$\delta\phi = 0.1058\frac{\delta t}{m}. \tag{3.71}$$

To find the angle $\Delta\phi$ moved through over one pass of the polar orbit we need the coordinate period of the orbit, T_θ. We can get this from the equations of motion in Section 3.10. From equations (3.46) and (3.44) we have

$$\frac{d\theta}{dt} = \frac{d\theta/d\lambda}{dt/d\lambda} = \frac{(Q_{ph} + m^2E_{ph}^2\cos^2\theta)^{1/2}\Delta}{AE_{ph}}. \tag{3.72}$$

Separating the variables, substituting for A and dividing through by E_{ph}^2 gives

$$\Delta\int_0^{T_\theta} dt = \int_0^{2\pi} [(r_{ph}^2 + m^2)^2 - m^2\Delta\sin^2\theta]\left[\frac{Q_{ph}}{E_{ph}^2} + m^2\cos^2\theta\right]^{-1/2} d\theta, \tag{3.73}$$

where $Q_{ph}/E_{ph}^2 = (11 + 8\sqrt{2})m^2$ for a polar photon orbit. (See problem 3.31.) Substituting for Q/E_{Ph}^2 and r_{ph} and evaluating the integral numerically, we get

$$2 \times 4.724 \times T_\theta/m = 283.5164 \tag{3.74}$$

and hence $T_\theta = 30.010m$. For comparison, the period of a photon orbit around a non-spinning black hole is $32.65m$, whilst the periods of photon orbits in the equatorial plane range from $12.57m$ for a co-rotating orbit to $43.99m$ for a contra-rotating one. So the spin of a black hole has only a small effect on the period of a polar orbit.

Substituting the value of T_θ into (3.71) gives $\Delta\phi = 3.175$ rad. So the node of each complete polar orbit is dragged through an angle just over π in the sense of rotation of the black hole.

The dragging of the orbital plane of a non-equatorial orbit is known as the *Lense–Thirring effect*. The Lense–Thirring effect has been measured to an accuracy of about 20% using the two laser ranging satellites LAGEOS and LAGEOS II in orbit about the Earth. Short laser pulses aimed at the satellites are reflected back by corner cubes giving the distance to an accuracy of a few millimetres. (See Ciufolini *et al.*, 2007.)

The mental picture that the term frame dragging conjures up is that of a whirlpool dragging everything around in the direction of its sense of rotation. This picture should be treated with caution as it can sometimes mislead us. As we have just seen the picture works for the case of polar orbits which are dragged around in the sense of the rotation of the hole. It also works for the case of a zero angular momentum particle in free fall in the equatorial plane (Section 3.8). On the other hand the whirlpool picture does not help us when we consider particles in circular orbits in the equatorial plane. We saw in Section 3.13.1 that for circular orbits having $r > 9m$ particles can orbit in either sense about the black hole. But surprisingly, for an orbit at a given radial coordinate, it is the particle orbiting in the opposite sense to the rotation of the hole that goes fastest. This can be seen from equation (3.60) and is the contrary of what the whirlpool analogy might lead us to expect.

Problem 3.31. *Show that*

$$\frac{Q}{E_{ph}^2} = (11 + 8\sqrt{2})m^2.$$

3.15 The ergosphere

Despite the fact that a black hole is shielded by an event horizon it is nevertheless possible to extract its rotational energy. This implies that the rotational energy is located outside the event horizon. Energy extraction is achieved by exploiting the properties of the region between the event horizon and the static limit surface of a Kerr black hole. This region has consequently been named the ergosphere (from the Greek for energy). This unexpected property of spinning black holes relies on the existence of negative energy orbits for particles inside the ergosphere. In this section we demonstrate the existence of these negative energy states. We then describe a method, the Penrose process, that in principle allows the extraction of energy.

3.15.1 *Negative energy orbits*

First some comments on energy. The specific energy of a particle having four-velocity u^μ as measured locally by an observer with four-velocity u^μ_{obs} is given by $u_\mu u^\mu_{obs}$. This can be a function of position but is always positive. The energy of this particle measured from infinity is $E = u_\mu k^\mu$, where $k^\mu = (1,0,0,0)$ in Boyer–Lindquist coordinates. This energy is a constant for a free particle, independent of position, but need not be positive. Recall that a negative energy $E < 0$, simply means that it takes more than the rest mass energy of the particle to remove it to infinity.

Returning to the equation for radial motion (3.25), we can consider this as a quadratic equation for E:

$$\rho^4 \left(\frac{dr}{d\tau}\right)^2 = r^2 \left(r^2 + a^2 + \frac{2ma^2}{r}\right) E^2$$

$$- 4maEL_z - r^2 \left(1 - \frac{2m}{r}\right) L_z^2 - \Delta(Q + r^2). \quad (3.75)$$

Confining ourselves to motion in the equatorial plane, we can solve this quadratic for E even in the case that $dr/d\tau \neq 0$. This gives

$$E = \frac{2amLr \pm \sqrt{4a^2m^2L^2r^2 - [r^4 + a^2(r^2 + 2mr)][L^2a^2 - (r^2 + L^2)\Delta - \dot{r}^2r^4]}}{r^4 + a^2(r^2 + 2mr)},$$

$$(3.76)$$

where we have written $\dot{r}$ for $dr/d\tau$ and L for L_z. From equation (3.15) we see that to keep $dt/d\tau > 0$ (so that the velocity four-vector points to the future) we must have

$$E > \frac{2marL}{r^4 + a^2(r^2 + 2mr)}.$$

So we take the positive square root in (3.76). If $L < 0$ then we have the potential for E to take negative values, provided the contribution from the square root can be kept small enough. The terms under the square root can be rearranged to collect those involving L^2:

$$4a^2m^2L^2r^2 - [r^4 + a^2(r^2 + 2mr)][L^2a^2 - (r^2 + L^2)\Delta - \dot{r}^2r^4]$$

$$= r^4\Delta L^2 + [r^4 + a^2(r^2 + 2mr)](r^2\Delta + \dot{r}^2r^4). \tag{3.77}$$

The terms that do not involve L^2 can be made small by restricting the component of motion in the radial direction, $\dot{r}$, and by taking $r^2 \ll L^2$. Since L is the angular momentum per unit mass this last condition can always be satisfied by taking the mass of the particle small enough. So consider orbits for which these terms can be neglected. Now, at the static limit surface in the equatorial plane we have $r = r_{st} = 2m$ and $\Delta = a^2$, so inside r_{st} the term $r^4\Delta L^2$ always contributes less than $(2m)^2r^2a^2L^2$, and hence its square root always contributes less than $+2am|L|r$ to E. Thus,

$$E < (2amLr + 2am|L|r)/[r^4 + a^2(r^2 + 2mr)]$$

and if $L < 0$ then $E < 0$. So there are orbits inside r_{st} with $L < 0$ for which also $E < 0$, or, in words, within the static limit surface there are retrograde orbits which have negative energy. As we said, this means that the energy required to remove a particle in such an orbit to infinity is greater than its rest mass.

Problem 3.32. *By finding $k_\mu k^\mu$ show that the 4-vector $(k^\mu) = (1, 0, 0, 0)$ is spacelike inside the static limit surface.*

Using a standard result in special relativity that the scalar product of a spacelike vector with a timelike vector can be negative we deduce from problem 3.32 that the energy $E = u_\mu k^\mu$ of a particle with 4-velocity u^μ (necessarily timelike) can be negative. The following problem addresses the direct proof of this result for the special case of a photon by explicit calculation.

Problem 3.33. *For a photon in an azimuthal direction in the equatorial plane use equation (3.48) for Ω, the definition of the angular velocity of zero* *angular momentum orbits, $\omega = -g_{03}/g_{33}$, and the relation $p_\mu = g_{\mu\nu}p^\nu$ to show that*

$$E_{ph} = \frac{dt}{d\lambda}(g_{00} + g_{03}\Omega)$$

and hence that

$$E_{ph} = \frac{2amL_{ph}r + |L_{ph}|r^2\Delta^{1/2}}{A}.$$

Demonstrate that this is consistent with (3.76) with $\dot{r} = 0$ and with $m_0 = 0$ for a photon. Deduce that for a counter-rotating particle in the ergosphere (where $r^2\Delta^{1/2} < 2amr$) E_{ph} can be negative if L_{ph} is negative.

Note that although the conserved energy per unit mass relative to an observer at infinity E can be negative, the local energy per unit mass γ is always positive.

3.15.2 *Energy and angular momentum*

We shall show that the angular momentum per unit mass $L_n = |L|$ and energy per unit mass $E_n = |E|$ of a negative energy particle satisfies the relation $E_n \leq \omega L_n$, a result we shall need in Chapter 4. Starting from the relation

$$g^{\mu\nu}u_\mu u_\nu = 1,$$

we solve the quadratic in u_0 to get

$$u_0 = -\frac{g^{03}}{g^{00}}u_3 \pm \frac{\left[(g^{03})^2 u_3^2 - g^{00}\left(g^{11}(u_1)^2 + g^{22}(u_2)^2 + g^{33}(u_3)^2 - 1\right)\right]^{1/2}}{g^{00}},$$

which becomes

$$u_0 = -\omega u_3 \pm \frac{\Delta^{1/2}}{A}\left[\frac{\rho^4}{\sin^2\theta}(u_3)^2 + A\Delta(u_1)^2 + A(u_2)^2 + \rho^2 A\right]^{1/2}. \quad (3.78)$$

On the horizon we have $\Delta = 0$ so

$$u_0 = -\omega_+ u_3.$$

As $u_0 = E$ and $u_3 = -L$ this gives

$$E = \omega_+ L.$$

So on the horizon positive L gives positive E, $L = 0$ gives $E = 0$ and negative L gives negative E. Away from the horizon we must take the positive square root in equation (3.78) to ensure that u_0 is positive for zero angular momentum, $u_3 = 0$. Then, in all cases, $u_0 + \omega u_3 \geq 0$ or

$$E - \omega L \geq 0.$$

For negative angular momentum particles, with $L = -L_n$ and $E = -E_n$ both negative, this reads $E_n \leq \omega L_n$.

3.15.3 *The Penrose process*

Consider a particle of energy per unit mass E that decays into two particles somewhere in the ergosphere where negative energy orbits are possible. Assume further that one of the decay products has positive angular momentum and positive energy E_p, while the other product has both negative angular momentum and negative energy E_n. So $E = E_p + E_n$ and $E_p > E$. Let the negative energy particle be captured by the hole, thus reducing its mass and angular momentum. The positive energy particle escapes to infinity with more than the initial energy E. As seen by the outside world therefore, this process extracts energy (and angular momentum) from the hole.

3.15.4 *Realising the Penrose process*

Despite its simplicity, the Penrose process is unlikely to provide a realistic mechanism for the extraction of energy from a black hole. To explain this statement consider the following. The energy per unit mass of a particle in the last stable circular orbit of an extreme Kerr black hole ($E = 1/\sqrt{3}$) is the lowest achievable energy for an orbiting particle entering the ergosphere from the outside. To realise the Penrose process this particle must decay into two fragments, one of which must be projected against the rotation of the hole into a negative energy orbit. The question that arises is, what is the size of the velocity boost required to do this? To answer this, we need to find the velocity with respect to a local ZAMO of a particle in the last stable circular orbit.

According to equation (3.37) the 4-velocity of a ZAMO in the equatorial plane is

$$(u^\mu_{ZAMO}) = \left(\left(-\frac{g_{33}}{\Delta} \right)^{1/2}, 0, 0, \omega \left(-\frac{g_{33}}{\Delta} \right)^{1/2} \right). \tag{3.79}$$

The 4-velocity of a particle in a circular orbit has the form

$$(u_\mu) = (u_0, 0, 0, u_3) = (E, 0, 0, \mp L).$$

Therefore the energy per unit mass in the ZAMO frame is

$$u^{\mu}_{ZAMO}u_{\mu} = \left(\frac{-g_{33}}{\Delta}\right)^{1/2}(u_0 + \omega u_3),$$
$$= \gamma,$$

where γ is the Lorentz gamma-factor of the particle with respect to a ZAMO.

We now write the expression for the energy in the ZAMO frame in terms of u_3 by using equation (3.78). For motion in the equatorial plane with r constant this gives

$$u_0 + \omega u_3 = \left(r^2 u_3^2 + A\right)^{1/2}\frac{r\Delta^{1/2}}{A}$$

and hence

$$\gamma = \left(\frac{-g_{33}}{\Delta}\right)^{1/2}\left(r^2 u_3^2 + A\right)^{1/2}\frac{r\Delta^{1/2}}{A}.$$

We can solve this for the speed of the particle in the ZAMO frame:

$$v^2 = \frac{\gamma^2 - 1}{\gamma^2} = \frac{r^2 u_3^2}{r^2 u_3^2 + A}. \tag{3.80}$$

Now the angular momentum of a particle in a circular orbit is given by equation (3.56). Multiplying numerator and denominator by r^2 this is

$$L = \pm\frac{m^{1/2}\left(r^2 \mp 2am^{1/2}r^{1/2} + a^2\right)}{r^{3/4}\left(r^{3/2} - 3mr^{1/2} \pm 2am^{1/2}\right)^{1/2}}, \tag{3.81}$$

where the upper sign refers to prograde orbits and the lower to retrograde orbits. Now substitute equation (3.81) into equation (3.80). After some algebra this gives

$$v = \pm\frac{m^{1/2}\left(r^2 \mp 2am^{1/2}r^{1/2} + a^2\right)}{\Delta^{1/2}\left(r^{3/2} \pm am^{1/2}\right)}. \tag{3.82}$$

We now consider various cases.

Case (1) A non-rotating hole ($a = 0$). In this case

$$v = \pm\frac{m^{1/2}}{r^{1/2}(1 - 2m/r)^{1/2}}.$$

For a particle in the last stable orbit, $r_{ms} = 6m$, we get $v = 1/2$ (or, in physical units, $v = c/2$).

Case (2) Retrograde orbits. For $a = m$ (the extreme Kerr case) the last stable retrograde circular orbit is at $r = 9m$ (Section 3.13.1) for which equatin (3.82) gives $v = -11c/26$ (in physical units). However, $r = 9m$ is outside the static limit so the energy is positive. Thus there are no retrograde circular orbits of negative energy.

Case (3) Prograde orbits. For $a = m$ the last stable prograde circular orbit is at $r = m$. To find the corresponding velocity from equation (3.82) we must take the limit $r \to m$ (since direct substitution gives an indeterminate form 0/0). Putting $r = m + \delta$ and letting $\delta \to 0$ we get $v = 1/2$ (or $c/2$ in physical units). The energy of a particle in this orbit was shown to be $1/\sqrt{3}$ in Section 3.13.3.

So now we know the speed of a particle in the last stable circular orbit which we are hoping will break up to yield a particle of negative energy. Consider the limiting case of a zero energy decay product. What is the speed of this particle as seen by a ZAMO? We saw in Section 3.9.1 that a zero energy particle at $r = m$ is at rest with respect to a ZAMO.

Evidently a particle in the last stable orbit of an extreme Kerr black hole requires a velocity boost of $c/2$ to reduce its energy from $E = 1/\sqrt{3}$ to $E = 0$. To achieve a negative energy trajectory requires an even greater velocity boost. Bardeen *et al.* (1972) conclude that no plausible astrophysical process can lead to such a boost in velocity.

3.16 Spinning up a black hole

It might be thought that by allowing matter co-rotating with a black hole to be accreted the hole could be spun up indefinitely. We show below that in fact a black hole cannot be spun up beyond $a = m$ by adding mass. This is important because, as we shall see, holes with $a > m$ have some extremely undesirable properties. The following argument is from Bardeen (1970). (See also Lynden-Bell, 1978.)

We consider mass falling into the hole from the last stable circular orbit. Note that any attempt to increase the angular momentum of the hole by accreting matter from other than this orbit will not help. Projecting a particle with a radial component of velocity will carry in less angular momentum and adding angular momentum to the orbit will cause the particle to escape to infinity and not be accreted at all.

When a rest mass dm_0, having energy per unit mass E given by equation (3.58), falls into a black hole from the last stable circular orbit the black hole acquires an increment of mass dm obtained from equation (3.58)

$$dm = E dm_0 = dm_0 \left(1 - \frac{2}{3}\frac{m}{r}\right)^{1/2}, \tag{3.83}$$

with $r = r_{ms}$. It also acquires angular momentum per unit mass given by equation (3.61), so the change in the angular momentum $j = am$ of the black hole is

$$dj = L dm_0 = dm_0 \frac{2m}{3\sqrt{3}} \left[1 + 2\left(3\frac{r_{ms}}{m} - 2\right)^{1/2}\right] \tag{3.84}$$

$$= 2m \, dm \frac{x_{ms}^{1/2}}{3(3x_{ms} - 2)^{1/2}} \left[1 + 2\left(3x_{ms} - 2\right)^{1/2}\right], \tag{3.85}$$

where we have substituted for dm_0 from equation (3.83) and put $x_{ms} = r_{ms}/m$. We want to find how a/m changes so let $\alpha = a/m = j/m^2$ and return to equation (3.59), but now considered as an equation for α. Solving this quadratic in α gives

$$\alpha = \frac{1}{3}\left(x_{ms}\right)^{1/2} \left[\pm 4 \mp \left(3x_{ms} - 2\right)^{1/2}\right].$$

To check the signs consider a co-rotating particle around an extreme Kerr black hole. For $x_{ms} = 1$ we get $\alpha = 1$ which is correct. For a counter-rotating particle $x_{ms} = 9$, which gives $\alpha = 1$ as it should. We have $dj = d(\alpha m^2) = m^2 d\alpha + \alpha dm^2$, which, rearranging, gives

$$m^2 \frac{d\alpha}{dm^2} = \frac{dj}{dm^2} - \alpha \tag{3.86}$$

and using $dm^2 = 2m \, dm$ in equation (3.85), finally

$$m^2 \frac{d\alpha}{dm^2} = \frac{\left[1 + 2\left(3x_{ms} - 2\right)^{1/2}\right] x_{ms}^{1/2}}{3\left(3x_{ms} - 2\right)^{1/2}} - \alpha. \tag{3.87}$$

This is an implicit equation for the change in α where x_{ms} is a function of α from (3.59). It is sufficient for us to obtain our main result: if $a = m$, then $r_{ms} = m$, so $\alpha = 1$ and $x_{ms} = 1$ and we find $d\alpha/dm^2 = 0$. Thus a/m cannot increase beyond 1. The black hole can still accrete, but with $a = m$ always.

There is a simple relationship between r_{ms} and the black hole mass as it accretes. We can eliminate α from (3.87) in favour of x_{ms} by writing

$$m^2 \frac{d\alpha}{dm^2} = m^2 \frac{d\alpha}{dx_{ms}} \frac{dx_{ms}}{dm^2}$$

and substituting for α and $d\alpha/dx_{rm}$. This gives

$$\frac{m^2}{x_{ms}} \frac{dx_{ms}}{dm^2} = -1,$$

from which

$$x_{ms} m^2 = \text{constant}. \tag{3.88}$$

Thus r_{ms} changes inversely with m. Note that equation (3.88) applies to accretion from both co-rotating and contra-rotating orbits.

Problem 3.34. *An extreme black hole continues to accrete from the last stable orbit, $r_{ms} = m$. Show that it remains an extreme Kerr hole with $j = m^2$.*

3.16.1 *From Schwarzschild to extreme Kerr black hole*

We want to look next at what happens when we start from a Schwarzschild black hole of mass m_S and spin it up by accretion of matter from the last stable circular orbit to an extreme Kerr black hole of mass m_K. The radius of the last stable orbit initially is $r_{ms} = 6m_S$ and is m_K finally. Using (3.88) we get $6m_S^2 = m_K^2$, so $m_K = \sqrt{6}m_S$ and the black hole mass has increased by $(\sqrt{6} - 1)m_S$ or 145%. This is not the same as the quantity of *rest mass* that has been added. To get this we use (3.83) in the form

$$dm_0 = \frac{dm}{\left(1 - \frac{2}{3}\frac{m}{r_{ms}}\right)^{1/2}}.$$

To do the integral we have to remember that r_{ms} is a function of m; in fact, from (3.88) again, $r_{ms} = 6m_S^2/m$. Thus

$$\Delta m_0 = \int_{m_S}^{\sqrt{6}m_S} \frac{dm}{\left(1 - \frac{1}{9}\frac{m^2}{m_s^2}\right)^{1/2}}$$

$$= 3m_s \left[\sin^{-1}\left(\frac{2}{3}\right) - \sin^{-1}\left(\frac{1}{3}\right) \right],$$

or $\Delta m_0 = 1.85 m_S$. But we have shown that the black hole mass has increased by just $1.45 m_S$. Thus $0.4 m_S$ or 22% of the rest mass has been radiated away in the process of accretion.

Note that the possibility of spinning up a black hole to the extreme Kerr limit depends on one crucial assumption, namely that we can neglect the effect of the radiation emitted by the accreting matter on the growth of the black hole. This is not the case. Some radiation will inevitably fall into the hole and add to its mass and angular momentum. From Section 3.10.3 we see that the capture of radiation that reduces the angular momentum (counter-rotating photons) is more likely than the capture of positive angular momentum radiation. The upshot of this is that a black hole cannot be spun up to $a = m$. (In fact the limit is $a = 0.998m$ (Thorne, 1974).) This outcome is comforting for two reasons. We shall see later that the structure of the spacetime of an extremal Kerr black hole is different from that of the standard Kerr model with $a < m$. Thus, in order to create an extremal hole, the global spacetime structure would have to be changed discontinuously, which is intuitively unphysical. The second reason, which will appear in Section 4.4.2, is that the result that a must remain less than m generalises to a statement of the third law of thermodynamics for black holes.

Problem 3.35. *An extreme Kerr black hole can be spun down to a Schwarzschild black hole by accretion of contra-rotating particles from the last stable circular orbit. Show that this will increase the black hole mass by 22.5% and that 4.7% of the accreted rest mass is radiated away in the process.*

Problem 3.36. *Matter is supplied at a constant rate to an initial Schwarzschild black hole to spin it up to near extreme Kerr ($a = 0.998m$, say). By what ratio does the luminosity of the accretion disc increase?*

3.17 Other coordinates

The singularity in the metric at $\Delta = 0$ is a coordinate singularity not a physical one, just like the Schwarzschild singularity at $r = 2m$. As in the Schwarzschild case the easiest way to see this is to transform to coordinates in which the metric is non-singular. The coordinates analogous to the Eddington–Finkelstein coordinates for Schwarzschild are the ingoing Kerr coordinates $(v, r, \theta, \tilde{\phi})$ and the outgoing coordinates $(u, r, \theta, \tilde{\phi})$. The ingoing Kerr coordinates are defined by

$$dv = dt + (r^2 + a^2)\frac{dr}{\Delta},$$

$$d\tilde{\phi} = d\phi - a\frac{dr}{\Delta}.$$

With these coordinates the metric becomes

$$ds^2 = \left(1 - \frac{2mr}{\rho^2}\right) dv^2 - 2dr\,dv - \rho^2 d\theta^2 - A\frac{\sin^2\theta}{\rho^2}d\tilde{\phi}^2$$

$$+2a\sin^2\theta d\tilde{\phi}dr + \frac{4amr}{\rho^2}\sin^2\theta d\tilde{\phi}dv,$$

which is clearly not singular at $\Delta = 0$.

Note that the transformation of t causes coordinate time v to pass increasingly quickly as we approach the horizon, which, just as in Schwarzschild, counteracts the 'stalling' of the infalling trajectory there. But in addition the transformation in ϕ 'unwinds' the angular coordinate increasingly rapidly as the horizon is approached to counteract the infinite winding of the infalling trajectory that we found in Section 3.8.1. One further difference from the Schwarzschild case is that the lack of spherical symmetry means that the picture we draw, with the angle θ suppressed, is different for different values of θ. Note also however, that most geodesics (and hence light rays) do not have constant values of θ, so in general a figure with θ suppressed does not give a very realistic view of the spacetime. (It does not allow one to picture the free fall particle paths.) Exceptions occur in the equatorial plane ($\theta = \pi/2$) and for polar orbits.

As for Schwarzschild, to draw the spacetime diagram we define a new time $\tilde{t} = v - r$, and plot the trajectories of light rays in $(\tilde{t}, r, \tilde{\phi})$ space. This is shown in Fig. 3.4 for the case $\theta = \pi/2$. It is clear from the diagram why we get two special null surfaces: on the static limit surface the counter-rotating null lines have been 'unwound' by the rotation of the hole as the 'straight' null lines that generate the surface, while on the horizon the outgoing light rays are wound up to remain in the surface.

3.18 Penrose–Carter diagram

In the case of a Schwarzschild black hole we obtained the continuation of the geometry into the interior of the horizon by transforming to Kruskal coordinates. We can do a similar thing for the Kerr solution. It turns out that there are three very different cases to consider. There is also an important difference in relating the pure vacuum solutions to the physical collapse of a star, but we shall come to that in Section 3.18.1. The three cases are $a > m$, which will turn out to be the simplest, but is quite unphysical in interesting ways; $a = m$, the (unattainable) limiting case referred to as the extreme (or, sometimes, maximal) Kerr solution; and

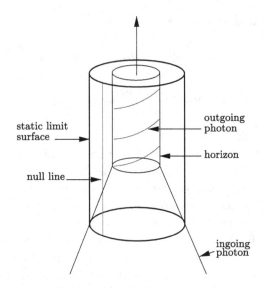

Fig. 3.4   Photons around a Kerr black hole (based on Frolov and Novikov, 1998, p. 79).

the apparently most complicated case $a < m$. (The following discussion is adapted from Townsend (unpublished).)

(i) $a > m$. This case is simple enough for us to provide the algebraic details. The metric has a coordinate singularity when $\Delta = (r - r_+)(r - r_-) = 0$, where $r_\pm = m \pm (m^2 - a^2)^{1/2}$. Thus for $a > m$, $\Delta = 0$ has no real solutions and there are no coordinate singularities (except the usual one in spherical polar coordinates at $\theta = 0$). It can be shown that there is a real singularity at $\rho = 0$, i.e. at $r = 0$, $\theta = \pi/2$. To see the structure of the spacetime in this case we can use a further set of coordinates, the Kerr–Schild coordinates (T, X, Y, Z). These are defined by

$$X = (r \cos \widetilde{\phi} + a \sin \widetilde{\phi}) \sin \theta$$

$$Y = (r \sin \widetilde{\phi} - a \cos \widetilde{\phi}) \sin \theta$$

$$Z = r \cos \theta$$

$$dT = dt - \left(\frac{r^2 + a^2}{\Delta} - 1 \right) dr,$$

where $\widetilde{\phi}$ is, as defined above,

$$\widetilde{\phi} = \phi - \int a \frac{dr}{\Delta},$$

and it follows that r is given implicitly as a function of (X, Y, Z) by

$$r^4 - (X^2 + Y^2 + Z^2 - a^2)r^2 - a^2 Z^2 = 0. \qquad (3.89)$$

The metric in Kerr–Schild coordinates becomes

$$ds^2 = dT^2 - dX^2 - dY^2 - dZ^2$$
$$- \frac{2mr^3}{r^4 + a^2 Z^2} \left[\frac{-r(X dX + Y dY) - a(X dY - Y dX)}{r^2 + a^2} - \frac{Z dZ}{r} + dT \right]^2.$$

At $r = 0$ we have $Z = 0$. Also if $r = 0$, then $dr = 0$, and if $\theta = \pi/2$ we get $X = a \sin \tilde{\phi}$, $Y = -a \cos \tilde{\phi}$, so $X^2 + Y^2 = a^2$. Thus the metric is singular on this ring in the (X, Y) plane (i.e. the equatorial plane). All inward null rays in the equatorial plane must intersect this ring.

To construct a picture of the spacetime for $\theta = \pi/2$ we imagine a further coordinate transformation that has no effect near $r = 0$ but brings the asymptotically flat region at large r in to finite values, and make a conformal transformation to a metric that is non-singular at the outer boundaries. This will result in the picture of Fig. 3.5. Since there is no horizon, light rays from the ring singularity escape to future null infinity, and the ring is visible to external observers. The ring is therefore a naked singularity. According to the cosmic censorship hypothesis (Section 2.12.1) this should mean that the Kerr solution with $a > m$ is unphysical (i.e. the gravitational field cannot be created from physical matter).

Recall that we should not be misled by the choice of symbols for coordinates into ascribing conventional meanings to them in relativity.

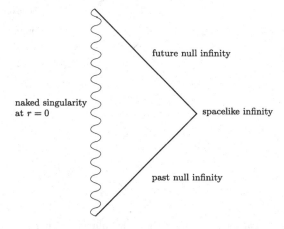

Fig. 3.5   Kerr spacetime for $a > m$ (based on Townsend, p. 81).

We have here another case in point: the metric gives a valid geometry for negative values of the coordinate r. The spacetime therefore continues through the disc $X^2 + Y^2 = a^2$ to another asymptotically flat region with negative r. The whole spacetime picture in the equatorial plane therefore contains another wedge to the left of the ring $r = 0$ in Fig. 3.5.

For other values of θ, $r = 0$ corresponds to points in the disc $X^2 + Y^2 < a^2$ interior to the ring. Null geodesics pass through the disc to the $r < 0$ region.

(ii) $a < m$. The analogy between the Kerr ingoing and outgoing coordinates and the two sets of Eddington–Finkelstein coordinates suggests that we can find the analogue of the Kruskal extension of Schwarzschild for the Kerr metric. This is indeed the case. It turns out that the algebraic details are not very illuminating (see, e.g. Frolov and Novikov, 1998, appendix) so we just give the result in the form of the Penrose–Carter diagram in the equatorial plane (Fig. 3.6). This contains the two

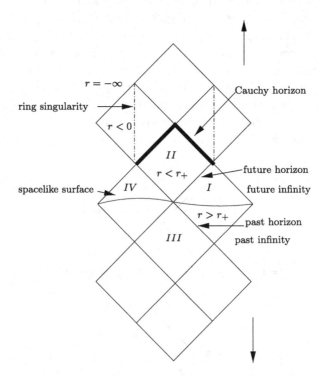

Fig. 3.6   Penrose–Carter diagram of the Kerr solution with $a < m$. The figure is repeated infinitely in both directions.

asymptotically flat regions I and IV, and the past and future horizons at
$r = r_+$, as we might expect from the analogy. The regions interior to the
horizons are quite surprising however. Regions II and III are bounded not
by a spacelike singularity (a singularity where time-like geodesics come to
an end, as in the Kruskal metric) but by the inner horizons at $r = r_-$. These
horizons act as limits to predictability (called Cauchy horizons) for exterior
observers. To understand this, consider that the positions and velocities of
all particles in the exterior of the black hole are given on a spacelike surface
in region I. The positions of these particles can be predicted up to the
Cauchy horizon. Beyond that a new particle could come in from another
region and collide with the given particles thereby upsetting any further
future predictions.

Within these Cauchy horizons are the ring singularities, which are
timelike singularities since they terminate spacelike geodesics. (Such singularities are avoidable in contrast to spacelike singularities.) There are also
further asymptotically flat regions with negative values of r, extending to
$r = -\infty$. And then the whole pattern is repeated indefinitely to the past and
to the future! Before commenting we look finally at the extreme Kerr case.

(iii) $a = m$. This is similar to $a < m$, except that the two horizons at
$r = r_\pm = m$ coincide, so the region of spacetime between them is no longer
present (Fig. 3.7).

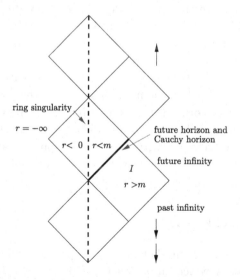

Fig. 3.7   Extremal Kerr spacetime.

Two comments are in order. The first is the question of what happens if we try to spin up a black hole by adding angular momentum. If the case $a > m$ is unphysical, there must be some limit to the accretion of angular momentum. We explored this in Section 3.16 and found that there is a limit. The second is the nature of the interior solution, and especially the possibility of exploring all those infinity of separate regions ('universes') within the horizon of a spinning black hole. One should be slightly suspicious of the physical reality of this picture, since the multiplicity of universes would appear to be created as soon as a Schwarzschild hole acquires the tiniest amount of angular momentum. We shall deal with this (as far as possible) in the next section.

3.18.1 *Interior solutions and collapsing stars*

We have emphasised that the Kerr solution, unlike the Schwarzschild solution, is not the metric of the vacuum spacetime exterior to a material body. Furthermore, in the collapse of a spherical body, the metric outside the body, both outside *and inside* the horizon, is exactly the Schwarzschild solution. Real stars are not exactly spherical, but if they are non-rotating the Schwarzschild solution is an increasingly good approximation to the metric as the collapse proceeds. This is not the case for the collapse of a rotating body: inside the horizon the metric is not even approximately the Kerr solution because the Kerr metric does not incorporate the gravitational waves that would be produced in a real collapse. This means that the exotic properties of the Kerr solution interior to the horizon do not represent the situation in a black hole that is formed by collapsing matter.

Suppose however that we were to be presented with an exact Kerr black hole. Could we explore the many universes of the interior? Probably not. Any observer exploring this metric represents a perturbation of the geometry. Near the inner horizon this perturbation cannot remain small, so the spacetime, including the observer, cannot be approximated by the exact Kerr black hole. In this sense the exact Kerr black hole interior cannot be explored without destroying it.

3.19 Closed timelike lines

The Kerr metric exhibits another interesting pathology. It contains closed timelike curves along which an observer would appear to be able to travel into his or her past. We show this first, and then discuss its significance.

Consider a path along which only ϕ is changing. The proper time along this path is given by

$$d\tau^2 = g_{33}d\phi^2 = \left[-a^2 \sin^2\theta \left(1 + \frac{r^2}{a^2} \right) - \frac{ma^2}{r} \left(\frac{2\sin^4\theta}{1 + \frac{a^2}{r^2}\cos^2\theta} \right) \right] d\phi^2.$$

Usually this quantity would be negative indicating a spacelike interval. We shall show there are curves along which this $d\tau^2$ is positive. Since curves in ϕ are closed, these will be closed timelike lines. Choose a path with $\theta = \pi/2 + \delta$ and $r = a\delta$, where δ is small. Then

$$\sin^2\left(\frac{\pi}{2} + \delta \right)(1 + \delta^2) \sim \cos^2\delta(1 + \delta^2) \sim 1$$

as $\delta \to 0$ and

$$\frac{a^2}{r} \left(\frac{2\sin^4\theta}{1 + \frac{a^2}{r^2}\cos^2\theta} \right) \sim \frac{a}{\delta} \left(\frac{2\cos^4\delta}{1 + \sin^2\delta/\delta^2} \right) \sim \frac{a}{\delta}$$

in the limit $\delta \to 0$. Putting this together we find

$$d\tau^2 \sim \left[-a^2 - \frac{ma}{\delta} \right] d\phi^2.$$

If δ is negative and sufficiently small then clearly $d\tau^2$ is positive as promised. If δ is negative, then $r = a\delta$ is negative. Thus, in the region $r < 0$ there are closed timelike paths. (Note that these are not geodesics, so some effort is required to navigate them.)

Problem 3.37. *The existence of closed timelike lines can also be demonstrated in the Kerr–Schild coordinates. Show this by calculating the element of proper time along a circle $r = a\delta$, $Z = $ constant, at constant time T. (Hint: along a circle in the (X, Y) plane $dX \propto -Y$, $dY \propto X$.)*

The problems with being able to travel into the past are obvious and potentially gruesome. (Add this to the Oedipus legend, for example.) Fortunately in this case the cosmic censorship hypothesis (if true) would prevent such a thing. In fact, as we have seen, we do not know how to make a Kerr black hole with $a > m$ using normal matter. On the other hand, travel into the past through singularity-free wormholes does appear to be possible (in principle) according to current theories (see Chapter 5). Whether more complete understanding of physics (including quantum gravity for example) will turn out in practice to allow only consistent time-travel or to prevent it altogether remains to be seen.

3.20 Charged black holes

Generally speaking astronomical bodies are electrically neutral, but there are several reasons why the addition of electric charge to black holes has been considered. The first is to add to the bank of exact solutions that can have their properties explored. In this respect charged (non-rotating) black holes act as a bridge to the study of Kerr black holes with which they have in common a Cauchy horizon and a timelike singularity, but lack some of the technical complexity. The second is one of completeness: the only external parameters black holes can have are mass, angular momentum and electrical charge (neglecting a hypothetical magnetic charge), so the Kerr–Newman metric of a charged spinning black hole is the most general black hole solution. The lack of other conserved quantities beyond mass, charge and angular momentum, associated with black holes is often expressed by saying that 'black holes have no hair'. One other importance that the extremal charged black hole has acquired in recent times is that it provides a connection to string theory.

The special case of a spherical charged black hole is known as the Reissner–Nordström solution. This has the metric

$$d\tau^2 = \left(1 - \frac{2m}{r} + \frac{q^2}{r^2}\right) dt^2 - \frac{dr^2}{\left(1 - \frac{2m}{r} + \frac{q^2}{r^2}\right)} - r^2 d\Omega^2.$$

Now $g_{00} = 0$ when

$$r_\pm = m \pm (m^2 - q^2)^{1/2}.$$

If $|q| > m$ then the roots of the quadratic are not real so g_{00} stays positive at all values of r except at $r = 0$, where there is a singularity.

The more interesting case is when $|q| < m$. There are three regions: (I) $r_+ < r < \infty$; (II) $r_- < r < r_+$; (III) $0 < r < r_-$.

In region (I) g_{00} is positive so it is possible to remain stationary.

In region (II) let $r = r_+ - \delta$. Then

$$g_{00} = \left[1 - \frac{2m}{r_+}\left(1 - \frac{\delta}{r_+}\right)^{-1} + \frac{q^2}{r_+^2}\left(1 - \frac{\delta}{r_+}\right)^{-2}\right] = -\frac{2\delta}{r_+^2}\left(m - \frac{q^2}{r_+}\right).$$

Now, $q < m$ and $r_+ > m$ so $g_{00} < 0$ in (II). Thus in region (II) a stationary particle would have a spacelike world line so rest is not possible. From the figure we see that particles can cross $r = r_+$ in one direction only. Thus $r = r_+$ is a horizon.

In region (III), $r = r_- - \delta$ and

$$g_{00} = -\frac{2\delta}{r_-^2}\left(m - \frac{q^2}{r_-}\right).$$

But now $r_- < m$ and $q^2 < m^2$ so $g_{00} > 0$. Figure 3.9 shows a plot of g_{00} against r.

In region (III) it is possible to remain stationary and avoid the singularity or to move into another region (see Fig. 3.8). The fate of a particle falling into a charged black hole is more interesting than that of

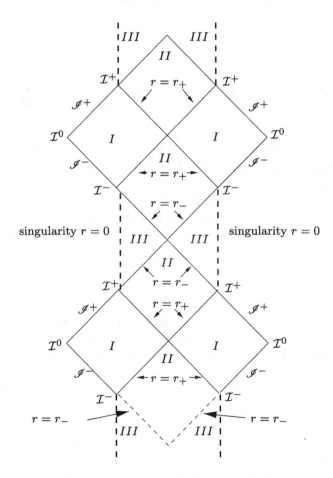

Fig. 3.8   Penrose–Carter diagram for the Riessner–Nordström solution with $q < m$.

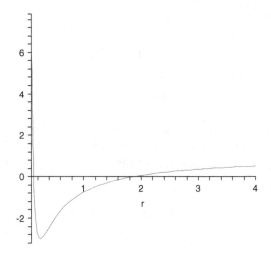

Fig. 3.9 g_{00} versus radius r in the case $m = 1$, $q = 0.5$. Note that within the inner horizon g_{00} becomes positive so the time coordinate is timelike as is the singularity at $r = 0$.

one falling into a Schwarzschild hole! The singularity is timelike, in contrast to the spacelike singularity in the Schwarzschild metric, which cannot be avoided.

Finally, for $q = m$, (the extremal case) we have $r_+ = r_- = m$ and $g_{00} = 0$ at $r = r_\pm$. There is no region (II) in this case.

The Kerr–Newman metric in Boyer–Lindquist coordinates for a spinning black hole with charge q is obtained from the Kerr metric by the replacement

$$2mr \to 2mr - q^2.$$

The properties of this solution can be explored in a similar manner to those of the Kerr metric. In particular the event horizon is given by $\Delta = r^2 + a^2 - 2mr + q^2 = 0$ and exists only for $m^2 \geq a^2 + q^2$.

Problem 3.38. *Show that the Kerr–Newman metric has an event horizon only in the case $m^2 \geq a^2 + q^2$.*

Chapter 4

Black Hole Thermodynamics

In this chapter we introduce the thermodynamics of black holes. Our strategy is to begin by establishing some mathematical identities for the Kerr solution (and, a fortiori, the Schwarzschild solution). These mathematical identities have a resemblance to the laws of thermodynamics if one assigns a temperature to a black hole. This resemblance was initially regarded as a curiosity, because a black hole did not appear to behave like a body with a temperature in one important respect: it did not radiate. In fact, the area theorem, which we introduce in Section 4.2, appears to prevent the emission of radiation from a Kerr black hole. However, once quantum theory is taken into account, Hawking showed that black holes do emit radiation precisely as required by the thermodynamical interpretation. Although it is beyond our scope to show this, these results can be extended to hold not just for Kerr black holes but for black holes in general. Thus black holes do obey the laws of thermodynamics.

4.1 Black hole mechanics

4.1.1 *Surface gravity*

In Section 2.8.2 we calculated the surface gravity of a Schwarzschild black hole by looking at the force on a rope supporting a body just above the horizon. In Kerr spacetime we cannot use this physical picture, because any part of the rope that was lowered into the ergosphere would be torn off by the rotation, since nothing can remain stationary inside the static limit surface. But we can obtain the acceleration of a zero angular momentum observer as seen from infinity in a similar way.

We begin by computing the local acceleration of a zero angular momentum observer. From equation (3.37) such an observer has the 4-velocity

$$u^\mu_{\text{ZAMO}} = \left[\left(\frac{-g_{33}}{\Delta \sin^2 \theta} \right)^{1/2}, 0, 0, \omega \left(\frac{-g_{33}}{\Delta \sin^2 \theta} \right)^{1/2} \right]$$

$$= \left[\left(\frac{A}{\rho^2 \Delta} \right)^{1/2}, 0, 0, \frac{2mra}{A} \left(\frac{A}{\rho^2 \Delta} \right)^{1/2} \right], \tag{4.1}$$

and hence the 4-acceleration

$$a^\mu = \frac{du^\mu_{\text{ZAMO}}}{d\tau} + \Gamma^\mu_{\nu\sigma} u^\nu_{\text{ZAMO}} u^\sigma_{\text{ZAMO}}. \tag{4.2}$$

The only non-zero contribution comes from the radial acceleration a^1. So equation (4.2) reduces to

$$a^1 = \left(u^0_{\text{ZAMO}} \right)^2 \left(\Gamma^1_{00} + 2\Gamma^1_{03}\omega + \Gamma^1_{33}\omega^2 \right),$$

where $u^3/u^0 = \omega$ as the observer has zero angular momentum. Substituting for ω from equation (3.28) gives

$$a^1 = \frac{\left(u^0_{\text{ZAMO}} \right)^2}{A^2} \left(A^2 \Gamma^1_{00} + 4mar A \Gamma^1_{03} + 4m^2 a^2 r^2 \Gamma^1_{33} \right). \tag{4.3}$$

We therefore need the components Γ^1_{00}, Γ^1_{30} and Γ^1_{33} of the affine connection. These can be calculated directly from equation (1.10) and the Boyer–Lindquist form of the metric. We obtain

$$\Gamma^1_{00} = -\frac{g^{11}m}{\rho^4} \left(r^2 - a^2 \cos^2 \theta \right),$$

$$\Gamma^1_{30} = \frac{g^{11}m}{\rho^4} a \sin^2 \theta \left(r^2 - a^2 \cos^2 \theta \right),$$

$$\Gamma^1_{33} = \frac{g^{11}}{2\rho^4} \sin^2 \theta \left(\rho^2 \frac{dA}{dr} - 2Ar \right).$$

From equation (1.13) the proper acceleration at coordinate radius r is given by

$$g^2_r = -a^\mu a_\mu = -g_{11} \left(a^1 \right)^2.$$

Substituting into equation (4.3) for the connection coefficients and taking out the common factor $-mg^{11}/\rho^4$ gives

$$g_r = \frac{m}{\rho^7 \Delta^{1/2} A} F$$

where

$$F = A^2 \left(r^2 - a^2 \cos^2 \theta\right) + 2ma^2 r \sin^2 \theta \left(2Aa^2 \cos^2 \theta - A'r\rho^2\right)$$

and $A' = dA/dr$. As we approach the horizon the locally measured acceleration $g_r \to \infty$. However, the acceleration g_∞ measured from infinity is finite since the two accelerations are related by a redshift factor as we explained in Section 2.8.2. Specifically

$$g_\infty = g_r \frac{E_\infty}{E_r}$$

(compare equation (2.38)). We compute $E_r/E_\infty = (A/\rho^2 \Delta)^{1/2}$ in the next section, giving

$$g_\infty = \frac{m}{\rho^6 A^{3/2}} F.$$

On the horizon, $\Delta = 0$, $r = r_+ = m + (m^2 - a^2)^{1/2}$, so

$$A = (r_+^2 + a^2)^2 = 4m^2 r_+^2.$$

Inserting these values into F gives, after some algebra,

$$F(r_+, \theta) = 4mr_+^2 (r_+ - m)\rho^6.$$

F is a function of θ through ρ, but the ρ^6 dependence will cancel in g_∞, leaving g_∞ independent of θ. Specifically, by definition, the acceleration on the horizon measured from infinity is $\kappa = g_\infty(r_+)$. So, using $(r_+ - m) = (m^2 - a^2)^{1/2}$, we get finally,

$$\kappa = \frac{(m^2 - a^2)^{1/2}}{2mr_+}. \tag{4.4}$$

In Schwarzschild spacetime the surface gravity must be constant on the horizon by virtue of the spherical symmetry. Our expression for surface gravity in the Kerr metric shows it to be constant on the horizon in this case also.

Note that for an extreme Kerr hole, $a = m$, we have $\kappa = 0$. This says that the surface gravity is zero because the rotation balances gravity. It is equivalent to a normal body rotating at break-up speed.

Problem 4.1. (i) *Show that the vector field* $(\chi^\mu) = (1,0,0,\omega)$ *is null on the horizon.* (ii) *More sophisticated treatments define the surface gravity in terms of the rate of change of the null vector field* $(\chi^\mu) = (1,0,0,\omega)$ *through*

$$\chi^\nu(\nabla_\nu \chi^\mu) = \kappa\chi^\mu.$$

Use the ingoing Eddington–Finkelstein coordinates (which are non-singular on the horizon) to derive the surface gravity of a Schwarzschild hole, $\kappa = 1/4m$, *from this definition.*

The result equation (4.4) for κ *can be obtained for a Kerr black hole using the ingoing Kerr coordinates introduced in Section 3.17.*

4.1.2 Redshift

To complete the derivation of the expression for surface gravity obtained in the last section we need to derive the expression for the redshift, at infinity, of light emitted from a zero angular momentum observer at radial coordinate r. The time component of the momentum 4-vector of a zero angular momentum photon is $p_0 = E_{ph}$ where E_{ph} is the conserved energy of the photon measured by an observer at infinity. Also $(L_{ph})_z = 0$ so $p_3 = 0$. Now we can obtain a relation between the energy of a photon measured locally at coordinate r by a ZAMO and the energy of the photon measured at infinity. This we do by evaluating the scalar product $p_\mu u^\mu_{\text{ZAMO}}$ in the local frame and in the original Kerr frame and equating the results. So

$$E_r = E_{\text{ph}} \left(\frac{-g_{33}}{\Delta \sin^2\theta} \right)^{1/2},$$

where, as usual, the components of u^μ_{ZAMO} in the Kerr frame are given by equation (4.1) and the components of $(p_\mu) = (E_{\text{ph}}, p_1, p_2, 0)$.

The conserved energy E_{ph} is the photon energy measured at infinity, $E_\infty = E_{\text{ph}}$ and therefore, using the identities in Section 3.3.1,

$$\frac{E_r}{E_\infty} = \left(\frac{A}{\rho^2 \Delta} \right)^{1/2}.$$

4.1.3 Conservation of energy

We now look at the energy change, or equivalently the change in mass, of a Kerr black hole under an external perturbation, such as the accretion of an element of mass. This will allow us to derive a relation expressing the conservation of energy for the black hole (not to be confused with the conservation of energy of a test body moving in the Kerr metric).

Recall from Section 3.2.2 (equation (3.9)) that the area of the event horizon is

$$A_h = 8\pi m \left[m + (m^2 - a^2)^{1/2} \right]. \tag{4.5}$$

A change in A_h is related to a change in m and a by

$$\delta A_h = 8\pi[2m\delta m + \delta m(m^2 - a^2)^{1/2} + m(m^2 - a^2)^{-1/2}(m\delta m - a\delta a)], \tag{4.6}$$

or

$$\delta A_h = \frac{8\pi}{(m^2 - a^2)^{1/2}} \left[2mr_+\delta m - a^2\delta m - am\delta a \right].$$

The angular momentum of the black hole is $j = ma$, so $\delta j = a\delta m + m\delta a$ and hence

$$\delta A_h = \frac{16mr_+\pi}{(m^2 - a^2)^{1/2}} \left[\delta m - \omega_+\delta j \right], \tag{4.7}$$

where $\omega_+ = a/(2mr_+)$ is the angular velocity of the black hole. We can rewrite this as

$$\delta m = \frac{\kappa}{8\pi}\delta A_h + \omega_+\delta j, \tag{4.8}$$

where

$$\kappa = (m^2 - a^2)^{1/2}/(2mr_+)$$

is the surface gravity. Equation (4.8), which expresses the conservation of energy, is the first law of black hole mechanics.

Problem 4.2. *Smarr showed that equation (4.8) could be integrated to obtain*

$$m = \frac{\kappa}{4\pi}A_h + 2\omega_+ j.$$

Verify this for the Kerr metric by explicit calculation.

4.2 The area of a Kerr black hole horizon cannot decrease

We expect that the addition of a particle of mass δm with zero angular momentum to a black hole will increase the mass by δm, so according to equation (4.7), the area of the event horizon will increase on addition of mass. Conversely, if we could add angular momentum without adding mass then equation (4.7) would show that the area would decrease. We certainly could not add angular momentum without mass by firing particles into the hole, so we would need to be more inventive. We look first at what happens to the surface area if we accrete matter from the last stable orbit. We shall then investigate what happens if energy and angular momentum are extracted from a black hole by the Penrose process. In this case δm and δj are negative, so does the minus sign in equation (3.84) come to the rescue and lead to an increase in area? In this example we can show that indeed it does. These examples illustrate the general result that the area of the horizon cannot decrease.

4.2.1 *Area change by accretion*

We consider accretion from the last stable orbit of an extreme Kerr black hole. From equation (3.58) the accretion of a mass δm_0 from the last stable prograde orbit at $r_{ms} = m$ gives rise to a change in mass of the hole, given by equation (3.83)

$$\delta m = \delta m_0 \left(1 - \frac{2}{3}\frac{m}{r_{ms}}\right)^{1/2} = \frac{\delta m_0}{\sqrt{3}}.$$

From equation (3.84) the associated change in angular momentum is

$$\delta j = \delta m_0 \frac{2m}{3\sqrt{3}}\left[1 + 2\left(3\frac{r_{ms}}{m} - 2\right)^{1/2}\right].$$

With $r_{ms} = m$ this becomes

$$\delta j = \frac{2m\delta m_0}{\sqrt{3}}.$$

So, since $\omega_+ = a/(2mr_+) = 1/2m$ for $a = m$ and $r_+ = m$, we get

$$\delta m - \omega_+\delta j = 0.$$

Direct substitution in equation (4.7) for $a = m$ would give the indeterminate form $0/0$. Therefore we have to evaluate the limit $a \to m$ more carefully.

Let $\varepsilon = m - a$ be small and positive. Then

$$\delta m = \frac{\delta m_0}{\sqrt{3}} + O(\varepsilon) \quad \text{and} \quad \delta j = \frac{2m\delta m_0}{\sqrt{3}} + O(\varepsilon).$$

Now substitute for δm, δj, and ω_+ into equation (4.7). This gives

$$\delta A_{\mathrm{h}} = \frac{16\pi r_+ m}{(m^2 - a^2)^{1/2}} \left(\frac{\delta m_0}{\sqrt{3}} - \frac{a}{r_+} \frac{\delta m_0}{\sqrt{3}} + O(\varepsilon) \right).$$

But

$$r_+ = m + (m^2 - a^2)^{1/2}$$

so

$$r_+ - a = m - a + (m^2 - a^2)^{1/2} = \varepsilon + (m^2 - a^2)^{1/2}.$$

Using this result gives

$$\delta A_{\mathrm{h}} = \frac{16\pi m}{(m^2 - a^2)^{1/2}} \frac{\delta m_0}{\sqrt{3}} (r_+ - a + O(\varepsilon))$$

$$\rightarrow \frac{16\pi}{\sqrt{3}} m\delta m_0 > 0.$$

Thus the horizon area does indeed increase in this case.

Problem 4.3. *Show that accretion of a photon from the circular photon orbit at radius r_{ph} also increases the area of an extreme Kerr black hole.*

Problem 4.4. *In the extreme case of an off-centre collision two Schwarzschild black holes merge to form an extreme Kerr black hole. Show that in this limit no gravitational radiation is emitted.*

4.2.2 Area change produced by the Penrose process

In the Penrose process a particle of energy E_0 inside the ergosphere decays into a positive energy particle of energy E_P and a negative energy particle of energy E_N, where $E_0 = E_N + E_P$. The positive energy particle escapes to infinity and the negative energy particle is captured by the black hole with a change (loss) of mass (or energy)

$$\delta m = E_N. \tag{4.9}$$

Similarly, since the negative energy particle has negative angular momentum, the hole loses angular momentum

$$\delta j = L_N. \tag{4.10}$$

The desired result hinges on the fact that $|E_N| \le \omega_+|L_N|$, which we derived in Section 3.15.2. Using this result we get, from equation (4.9) and equation (4.10)

$$|\delta m| \le \omega_+|\delta j|.$$

From equation (4.7), remembering that both δj and δm are negative, the area law is

$$\delta A_{\rm h} = \frac{8\pi}{\kappa}(\delta m - \omega_+\delta j) \ge 0.$$

A transformation can be reversed only if $\delta A_{\rm h} = 0$; if $\delta A_{\rm h} > 0$ the transformation is obviously irreversible (because the inverse transformation would violate the area law).

4.2.3 *The area theorem*

We have demonstrated the increase in area for particular processes and specifically for a Kerr black hole. Hawking showed that the result holds within classical general relativity, under certain technical assumptions, for any process and any black hole (not just those in a vacuum). This result is known as the Hawking area theorem, or the second law of black hole mechanics:

No interaction whatsoever can result in a decrease in the total surface area of a (classical) black hole.

4.2.4 *Irreducible mass*

Equation (4.5) can be written as

$$m^2 = \frac{A_{\rm h}}{16\pi} + \frac{4\pi j^2}{A_{\rm h}}. \tag{4.11}$$

If we reduce j to zero reversibly then $\delta A_{\rm h} = 0$ giving the smallest value for m^2. This is called the irreducible mass $m_{\rm ir}$, so we have

$$m_{\rm ir}^2 = \frac{A_{\rm h}}{16\pi}$$

or

$$A_{\rm h} = 4\pi\left(\frac{2GM_{\rm ir}}{c^2}\right)^2,$$

analogous to the expression in the Schwarzschild case, where $M_{\rm ir}$ is the irreducible mass in physical units.

We have therefore

$$m^2 = m_{\text{ir}}^2 + \frac{4\pi j^2}{A_{\text{h}}}.$$

The second term on the right is the contribution to the mass (or energy) of the hole coming from the rotation. This is another version of the energy equation (or the Smarr formula or the first law). Evidently 'the Schwarzschild black hole is the ground state of the Kerr black hole'.

We shall look at some consequences of the 'area theorem' (that the area of a black hole can never decrease) below.

Problem 4.5. *Show that equation* (4.5) *can be written in the form of equation* (4.11).

Problem 4.6. *Show that for a slowly rotating black hole the rotational energy can be expressed as* $m_{rot} = \frac{1}{2}I\omega_+{}^2$ *where* I, *the moment of inertia, is* $4m^3$.

4.2.5 *Maximum energy extraction*

We can use the above results to obtain a limit to the amount of energy that can be extracted from a spinning black hole. This is $m - m_{\text{ir}}$. But

$$m_{\text{ir}}^2 = \frac{A_{\text{H}}}{16\pi} = \frac{m}{2}\left[m + (m^2 - a^2)^{1/2}\right].$$

Now m_{ir} is a minimum, hence $m - m_{\text{ir}}$ is a maximum when $a = m$, and therefore for an extreme Kerr black hole we get

$$m - m_{\text{ir}} = m - \frac{m}{\sqrt{2}}.$$

The maximum energy that can be extracted is thus 29% of the mass of the black hole.

Problem 4.7. *Show that the maximum energy that can be extracted from the merging of two Schwarzschild black holes of equal mass to form a single black hole is 29% of the original mass.*

Problem 4.8. *Two extreme Kerr black holes of the same mass and equal but opposite angular momentum coalesce reversibly to form a Schwarzschild black hole. How much energy is radiated away in the process?*

4.2.6 *Naked singularities*

So what happens if we try to add mass and angular momentum to an extreme Kerr black hole? Since the area cannot decrease, we have, from equation (4.6),

$$\delta A_h = 8\pi\delta m \left[m + (m^2 - a^2)^{1/2} \right]$$
$$+ 8\pi m \left[\delta m + (m^2 - a^2)^{-1/2}(m\delta m - a\delta a) \right] \geq 0.$$

This can be re-arranged to give

$$\left[2m(m^2 - a^2)^{1/2} + 2m^2 - a^2 \right] \delta m \geq ma\delta a.$$

For $a \to m$ this becomes

$$m\delta m \geq a\delta a,$$

hence, integrating, $m^2 \geq a^2$, or, equivalently, $a \leq m$. The area theorem prevents the spinning up of a black hole beyond the extremal state. In other words, it prevents the formation of a naked singularity by this means. However, it appears to allow the extreme Kerr black hole with $a = m$ to be reached. In fact, we have seen in Section 3.16.1 that, if we go beyond the accretion of test matter to take into account all the physics of the accretion process, then this is impossible. The inability to reach $a = m$ in a finite number of physical operations is an example of the third law of black hole thermodynamics (Section 4.4.2).

4.3 Scattering of waves

4.3.1 *Superradiance*

So far we have looked at the scattering of particles by a black hole. In order to approach the quantum theory of black holes it will be of interest to consider how waves are scattered by a black hole. We shall confine ourselves to a brief outline. Details can be found in Frolov and Novikov (1998) or Futterman *et al.* (1988) and the references therein.

To simplify the discussion we consider a (hypothetical) scalar wave $\psi(t, r, \theta, \phi)$ for a massless field, but the conclusions we reach apply equally to the electromagnetic field. The massless scalar wave satisfies the covariant form of the wave equation

$$\nabla_\mu \nabla^\mu \psi = 0. \tag{4.12}$$

As can be verified by explicit calculation using equation (1.10), this can be written as

$$(-g)^{-1/2} \partial_\mu \left[(-g)^{1/2} g^{\mu\nu} \partial_\nu \psi \right] = 0, \tag{4.13}$$

where, as usual, $\partial_\mu \equiv \partial/\partial x^\mu$ and $g = \det(g_{\mu\nu})$. By taking $g_{\mu\nu}$ to be those of the spherically symmetric Schwarzschild metric, the equation can be written explicitly in (t, r, θ, ϕ) coordinates

$$\frac{\partial^2 \psi}{\partial t^2} - \frac{2}{r} \left(1 - \frac{m}{r} \right) \left(1 - \frac{2m}{r} \right) \frac{\partial \psi}{\partial r} - \left(1 - \frac{2m}{r} \right)^2 \frac{\partial^2 \psi}{\partial r^2}$$
$$- \left(1 - \frac{2m}{r} \right) \frac{1}{r^2 \sin^2 \theta} \frac{\partial}{\partial \theta} \left(\sin \theta \frac{\partial \psi}{\partial \theta} \right) - \left(1 - \frac{2m}{r} \right) \frac{\partial^2 \psi}{\partial \phi^2} = 0.$$

As can be readily verified by substitution, this has separable solutions of the form,

$$\psi(t, r, \theta, \phi) = R_{\omega L}(r) Y_{LM}(\theta, \phi) e^{\pm i\omega t}, \tag{4.14}$$

where L, M and ω are separation constants. The $Y_{LM}(\theta, \phi)$ are the usual spherical harmonics satisfying the angular part of the wave equation, which has the same form as in flat spacetime. A general solution for ψ would be made up of a linear superposition of these separable solutions with arbitrary coefficients, as usual.

The radial function $R_{\omega L}(r)$ satisfies

$$\left(1 - \frac{2m}{r} \right)^2 \frac{d^2 R_{\omega L}}{dr^2} + \frac{2}{r} \left(1 - \frac{m}{r} \right) \left(1 - \frac{2m}{r} \right) \frac{dR_{\omega L}}{dr}$$
$$+ \left[\omega^2 - \left(1 - \frac{2m}{r} \right) \frac{l(l+1)}{r^2} \right] R_{\omega L} = 0. \tag{4.15}$$

We now introduce the function

$$u_{\omega L}(r) = r R_{\omega L}(r)$$

and the 'tortoise' coordinate, equation (2.47),

$$r_* = r + 2m \log \left(\frac{r}{2m} - 1 \right). \tag{4.16}$$

Since it is obvious from equation (4.15) that $u_{\omega L}(r)$ depends on the parameters ω and L we drop the subscripts for clarity. The function $u = u_{\omega L}(r)$ then satisfies the Regge–Wheeler equation

$$\frac{d^2 u}{dr_*^2} + \omega^2 u - V_L(r)u = 0, \qquad (4.17)$$

where V_L is the effective potential

$$V_L(r) = \left(1 - \frac{2m}{r}\right)\left[\frac{l(l+1)}{r^2} + \frac{2m}{r^3}\right],$$

which is implicitly a function of r_* through (4.16).

Note that $r_* \to +\infty$ as $r \to \infty$ and $r_* \to -\infty$ as $r \to 2m$, so r_* covers the exterior of the black hole. Note also that, since the effective potential tends to zero on the horizon and at infinity, the solutions of (4.17) at infinity and at the horizon have the simple form in terms of r_*

$$u \sim e^{\pm i\omega r_*}.$$

The simple behaviour in terms of r_*, rather than r, reflects the long range behaviour of the gravitational force.

Consider the solution of the wave equation in which no waves emerge from the black hole. This requires that we include in ψ only ingoing modes at $r_* \to -\infty$, hence these solutions are often called the 'in' modes. As in the standard case of the one-dimensional wave equation (for example for waves on a string), this requires that ψ be a function of $r_* + t$ (and not $r_* - t$). From the form of ψ in equation (4.14) we see that we must have

$$u \sim e^{-i\omega r_*} \quad \text{as } r_* \to -\infty. \qquad (4.18)$$

Since ω is positive by definition, from a quantum mechanical viewpoint we can also see that this wave function has negative momentum $-\hbar\omega$ (since $-i\hbar\partial u/\partial r_* = -\hbar\omega u$) and hence represents a wave travelling to smaller r (or r_*).

Just as for the standard scattering of waves in one dimension, having set the conditions at $r_* \to -\infty$ we can in principle calculate from the wave equation what happens as $r_* \to +\infty$. The ingoing wave will be scattered by the black hole such that part goes down the hole and part is reflected back to future infinity (Fig. 4.1). We expect to get a wave that has both

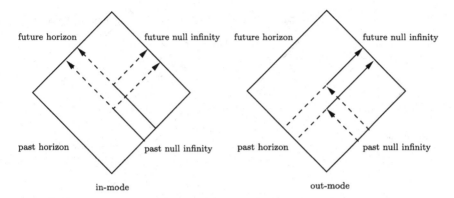

future horizon future null infinity future horizon future null infinity

past horizon past null infinity past horizon past null infinity

in-mode out-mode

Fig. 4.1   In-modes: rays from past infinity are scattered to future infinity and down the hole. Out-modes: incoming radiation from past infinity and the past horizon combine to ensure that there is no radiation crossing the event horizon (based on Frolov and Novikov, 1998, p. 93).

incoming and outgoing components:

$$u \sim a_{out}e^{i\omega r_*} + a_{in}e^{-i\omega r_*} \quad \text{as } r \to +\infty, \tag{4.19}$$

where a_{out} and a_{in} are functions of L and ω. It can be shown (problem 4.10) that these satisfy

$$1 + \left|a_{out}^2\right| = \left|a_{in}^2\right|. \tag{4.20}$$

We can define transmission and reflection coefficients by

$$T = \frac{1}{a_{in}} \quad \text{and} \quad R = \frac{a_{out}}{a_{in}},$$

whence equation (4.20) becomes

$$|T|^2 + |R|^2 = 1. \tag{4.21}$$

Values for the transmission coefficient (corresponding to a wave propagating into the black hole) and the reflection coefficient (corresponding to radiation scattered to infinity) can be found by numerical calculation, or approximately in special cases. For very high frequencies ($\omega \gg 1/m$) the ω^2 term in the Regge–Wheeler equation (4.17) is much larger than the effective potential V_L in the vicinity of $r = 2m$. Thus the potential term can be neglected and the waves are barely influenced by the hole. For very low frequencies ($\omega \ll 1/m$) the waves are almost entirely scattered back by the potential barrier in V_L. Figure (4.1) illustrates this solution (Frolov and Novikov, 1998).

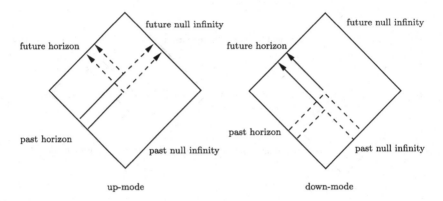

Fig. 4.2 The up- and down-modes.

Other solutions corresponding to simple boundary conditions are illustrated in Fig. 4.2. The complex conjugates of the in-modes are the out-modes (corresponding to no radiation falling down the hole). We can define also 'up-modes' which are purely outgoing at infinity

$$u \sim e^{i\omega r_*} \quad \text{as } r_* \to +\infty,$$

and the 'down-modes' which are the complex conjugates of the up-modes.

We have dealt with test particles and test fields in a fixed metric background, ignoring the effect that such additional energy would have on the gravitational field. This approximation will not affect any of our conclusions below. But we mention in passing that consideration of the effect on the metric of a Schwarzschild black hole by a material perturbation leads to an equation of the form (4.17), although with a different effective potential (Zerilli, 1970). The perturbed metric can be obtained from solutions of this equation. The results reveal a complex pattern of oscillations of the black hole referred to as quasi-normal modes. The details are beyond the scope of this book (Chandrasekhar, 1983).

Of more immediate relevance to the topic of this chapter is the extension of the treatment of wave scattering to rotating black holes. On inserting the Kerr metric coefficients in to equation (4.13), the scalar wave equation becomes

$$\left[\frac{(r^2 + a^2)^2}{\Delta} - a^2 \sin^2\theta\right] \frac{\partial^2\psi}{\partial t^2} + \frac{4mar}{\Delta} \frac{\partial^2\psi}{\partial t\partial\phi} + \left[\frac{a^2}{\Delta} - \frac{1}{\sin^2\theta}\right] \frac{\partial^2\psi}{\partial\phi^2}$$

$$- \frac{\partial}{\partial r}\left(\Delta \frac{\partial\psi}{\partial r}\right) - \frac{1}{\sin\theta} \frac{\partial}{\partial\theta}\left(\sin\theta \frac{\partial\psi}{\partial\theta}\right) = 0.$$

The equation can again be separated into normal modes as

$$\psi = R_{\omega LM}(r) S_{LM}(\theta) e^{iM\phi} e^{-i\omega t}.$$

We introduce an analogue of the coordinate r_*, in this case defined by

$$\frac{d}{dr_*} = \frac{\Delta}{r^2 + a^2} \frac{d}{dr},$$

and a new radial variable

$$u(r) = (r^2 + a^2)^{1/2} R_{\omega LM}(r),$$

where it is understood that $u(r)$ depends on the parameters ω, L and M, which we suppress for clarity. We obtain for $u(r)$ the radial equation

$$\frac{d^2 u}{dr_*^2} + \left\{ \frac{((r^2 + a^2)\omega - aM)^2 - \Delta(L(L+1) + a^2\omega^2 - 2aM\omega)}{(r^2 + a^2)^2} \right.$$
$$\left. - \frac{r^2 \Delta^2}{(r^2 + a^2)^4} - \frac{d}{dr_*}\left(\frac{r\Delta}{(r^2 + a^2)^2} \right) \right\} u = 0. \qquad (4.22)$$

This is the Teukolsky equation for a scalar wave; it reduces to the Regge–Wheeler equation when $a = 0$ (problem 4.9).

Now, we see from equation (4.22) that

$$u \sim e^{\pm i\omega r_*} \quad \text{as } r \to +\infty. \qquad (4.23)$$

At the horizon ($r_* \to -\infty$, $r = r_+$) we have $\Delta = 0$, which implies that $r_+^2 + a^2 = 2mr_+$ and hence

$$\frac{d^2 u}{dr_*^2} + \left(\omega - \frac{aM}{2mr_+} \right)^2 u \sim 0.$$

Thus

$$u \sim e^{\pm i(\omega - M\omega_+)r_*} \quad \text{as } r \to r_+ \qquad (4.24)$$

where $\omega_+ = a/(2mr_+)$ is the angular velocity of the hole, equation (3.29).

We can repeat the steps that led to equation (4.21). Now, however, we see that the transmitted wave (the one that goes down the hole, equation (4.24)) has a different wavenumber from the reflected wave (the component that is scattered to infinity, equation (4.23)). This is exactly analogous to the transmission and reflection of waves at the boundary between two different media: the transmission coefficient is multiplied by the ratio of wavenumbers such that

$$\frac{(\omega - M\omega_H)}{\omega} |T|^2 = 1 - |R|^2 \tag{4.25}$$

(problem 4.10). This leads to a novel feature that was not present in the spherical case. For wavemodes such that $\omega < M\omega_H$ the reflected amplitude exceeds that of the incidence wave (i.e. $R > 1$). Such waves are amplified by scattering, the energy coming from the rotation of the hole.

This amplification of waves by scattering is called superradiance. We can think of it as the induced emission of energy by the hole (induced by the incident radiation field). Recall that the induced emission of radiation from an excited atom in an electromagnetic field is a classical phenomenon. The effect of quantum mechanics is to cause spontaneous emission from the atom (emission independent of any external radiation field). We require both spontaneous and induced emission in order to achieve thermal equilibrium between an atom and a thermal radiation field. (Spontaneous and induced emission are described by the Einstein A and B coefficients, which must be related to each other to ensure thermal equilibrium with radiation having a Planck spectrum.) This prompts us to ask what might happen to a black hole in a thermal radiation bath. The surprise is not only that the inclusion of quantum theory does indeed lead to spontaneous emission from a rotating hole, but that, as we shall see later (Section 4.5.4), a Schwarzschild black hole is also predicted to radiate.

For completeness note that equation (4.22) generalises from a spin zero (scalar) field to fields of arbitrary spin s, and that in the general case it is called the Teukolsky equation. For example, spin $s = 1$ corresponds to the electromagnetic field, spin $s = 2$ to gravitational perturbations and spin $s = 1/2$ to (massless) neutrinos. It is a somewhat remarkable fact that this general wave equation in the Kerr background has separable solutions.

Problem 4.9. *Derive the asymptotic behaviour at large distances from the hole equation* (4.23) *and close to the horizon equation* (4.24) *of solutions of the scalar Teukolsky equation* (4.22). *Show that if we put $a = 0$ in the Teukolsky equation we recover the Regge–Wheeler equation.*

Problem 4.10. *For any two independent solutions of a second order linear differential equation, $u(r_*)$ and $v(r_*)$ (say) it can be shown that the quantity $W = u'v - v'u$ (where $'$ indicates d/dr_*) is constant. Evaluate W as $r_* \to -\infty$ by taking $u = u_{\omega L}(r_*)$ given by equation (4.18) and $v = u_{\omega L}^*$, the* *complex conjugate of u. Find W as $r_* \to +\infty$ by taking $u = u_{\omega L}(r_*)$ given by equation (4.19) and $v = u^*_{\omega L}$. Hence show that for the Regge–Wheeler equation (4.17) this leads to the relation (4.21) and for the Teukolsky equation (4.22) it leads to (4.25).*

4.4 Thermodynamics

The 'first law' of black hole mechanics equation (4.8) is rather reminiscent of the first law of thermodynamics: it says that the change in energy of a black hole δm comes partly from the change in rotational energy (the second term on the right) and partly from a term of the form $T\delta S$, provided that we identify (up to a constant of proportionality) the temperature T with the surface gravity κ and the entropy S with the surface area of the event horizon A_h. The constancy of κ on the horizon translates to the constancy of temperature between systems in equilibrium. And the area law, that the area of the horizon cannot decrease, is analogous to the second law of thermodynamics, that the entropy cannot decrease in a closed system. The analogy is consistent to the extent that this follows from the identification of the entropy of a black hole with its area through the first law. The problem with this analogy, in the context of classical general relativity, is that black holes are perfect absorbers but do not radiate: apparently, therefore, they cannot be assigned a 'real' temperature. This seemed to be an insuperable obstacle to making the formal analogy into a physical one.

The physical consistency of the picture is provided by the introduction of quantum mechanics. This raises the laws of black hole mechanics from mathematical analogies to physical descriptions through the Hawking process. According to this, black holes radiate like blackbodies at a temperature that depends on their mass (equation (4.28)).

In retrospect we can see why this should be so through various thought experiments, one of which we shall give below (following Kiefer, 1999). In effect these experiments make use of the fact that black holes can form parts of various physical systems and should not violate the thermodynamical behaviour of such systems. For example, if black holes have no entropy, it would be possible to reduce the entropy of a closed system containing a black hole by throwing things into the hole. In such circumstances, thermodynamics would not apply to systems containing black holes, for example, a star just past the point of collapse. (Incidentally, this situation is not rescued by the fact that an external observer never sees the final collapse: the entropy of a black hole, like its other attributes, relates to the properties of the spacetime on spacelike surfaces (such as $U + V =$ const. in Fig. 2.6 of Chapter 2), and not to what an external observer can see.)

There are numerous accounts of the Hawking process in the literature that attempt to explain what is happening using a simple quantum mechanical picture. These explanations are sometimes somewhat misleading. The Hawking radiation arises from a black hole that has been formed at a finite time in the past from the collapse of a star, and does not arise from an eternal black hole, as some of the simpler accounts might lead one to believe. Unless one arbitrarily imposes time-asymmetric boundary conditions, an eternal black hole is time-symmetric and therefore cannot radiate (since the emission of radiation would define a direction in time). Any picture that does not take the time-asymmetry of the collapse process into account cannot represent the quantum mechanics correctly. One well-known example of a dubious explanation is to imagine the quantum vacuum to be filled with short-lived particle-antiparticle pairs that come into existence for a time $\Delta E/h$ before annihilating again. In this time, one particle of negative energy falls into the hole and is captured while the other escapes to infinity as Hawking radiation. One can show that this picture leads to the correct order of magnitude for the blackbody temperature. However, any argument based on the uncertainty principle will lead to the correct functional form for the radiation on purely dimensional grounds. This does not prove the correctness of the physical picture. (For the sake of completeness we should add that the picture can be resurrected with the correct definition of particle and vacuum state, which are different for the collapsing star and for the eternal hole. The correlations in the eternal vacuum mean that the putative positive energy particle is cancelled out by interference with a negative energy particle from the past horizon, so there is no radiation. This, however, takes us beyond the scope of this book.)

In the next section we shall provide an argument that shows how the black hole temperature is required for consistency with thermodynamics. We shall then outline the quantum calculation which shows that black holes do have a temperature and hence that black holes do satisfy the laws of thermodynamics.

Note the parallel with Planck's original discovery of the quantum theory: here too classical physics was found to be inconsistent with thermodynamics, in this case with the thermodynamics of radiation, a situation that was resolved by the introduction of quantum theory.

We shall see that radiating black holes raise a number of fundamental issues which are unresolved. There are problems associated with trans-Planckian modes which we mention is Section 4.5.4 and with so-called firewalls (Section 4.7.2). The main issue, which we shall explain in Section 4.7.2 concerns the apparent loss of information by a radiating hole. For a time these problems cast doubt on Hawking's result. However, the result has since been re-derived in so many different ways in so many contexts that there seems to be no doubt that black holes radiate.

Problem 4.11. *Derive an expression for the Hawking temperature $T_h \propto \frac{hc^3}{GMk}$ by a dimensional argument, given that the temperature is inversely proportional to the black hole mass.*

4.4.1 Horizon temperature

We can show that it is necessary to associate a temperature with the horizon for consistency with thermodynamics (Keifer, 1999; Bekenstein, 1980). A box containing a mass m of thermal radiation at a temperature T is lowered from a distance towards a spherically symmetric black hole. For simplicity we assume that the mass of the walls of the box and of the rope can be neglected. As we saw in Section 2.3.8 (equation (2.23)) at a distance r from the black hole the energy in the box measured from infinity is given by

$$\mathcal{E}_r = m_0 \left(1 - \frac{2m}{r}\right)^{1/2}.$$

As the box approaches the horizon at $r = 2m$ the energy tends to zero, so the box gives up an energy m_0 in the process of being lowered from infinity.

Let the box now be opened and an amount of thermal radiation δm_0 be allowed to escape into the hole. This energy does not change the mass of the hole, as it is zero as seen from infinity. If the box is now raised, an energy of $m_0 - \delta m_0$ is expended to raise it back to infinity. So an energy δm_0 is dumped into the black hole, and at the end of the cycle δm_0 has been converted into external work at infinity. Therefore thermal radiation is being converted into work in a reversible cycle. The efficiency of the cycle is

$$\eta = \frac{\text{work out}}{\text{energy expended}} = \frac{\delta m_0}{\delta m_0} = 1,$$

which violates the second law of thermodynamics. Note that the sequence of operations does not violate the first law, but the violation of the second law would enable us to use the process to construct a perpetual motion machine of the second kind. In essence, disordered energy (radiation) is being converted to ordered work thereby decreasing the entropy of the universe. Since an energy δm_0 is dropped into the hole arbitrarily close to the horizon the black hole mass is not increased, so its area does not increase and neither therefore does its entropy. But the entropy of the box is lowered when it returns to infinity. Thus the reversible cycle leads to a decrease in the entropy in the world.

The resolution turns out to be to question how arbitrarily close the box can approach the horizon. In fact, the box must have a finite size since it must be larger than the wavelength of the enclosed radiation. This limits the distance to which the box can approach the hole. As a result we shall see that only part of the energy δm_0 can be converted into work.

Radiation at temperature T has its peak intensity at a wavelength

$$\lambda \sim \frac{hc}{kT} \tag{4.26}$$

according to Wien's law. This is where quantum theory comes in (as is evident from the appearance of Planck's constant in the formula). The box must have a proper length of at least λ. Therefore it can be lowered to within a coordinate distance l of the horizon, where

$$\lambda = \int_{2m}^{2m+l} \frac{dr}{(1 - 2m/r)^{1/2}}$$

$$= \int_{2m}^{2m+l} \frac{r^{1/2}dr}{(r - 2m)^{1/2}} \sim (2m)^{1/2} \int_0^l \frac{d\varepsilon}{\varepsilon^{1/2}} \sim 2\sqrt{2ml}.$$

Thus, rearranging,

$$l \sim \frac{\lambda^2}{8m}. \tag{4.27}$$

We now recalculate the energy of the box before and after opening it. Before opening it is

$$\mathcal{E}_r = m_0 \left[1 - \frac{2m}{(r + l)} \right]^{1/2} \sim m_0 \left[1 - \frac{2m}{(2m + l)} \right]^{1/2}$$

$$\sim m_0 \left(\frac{l}{2m} \right)^{1/2} \sim m_0 \frac{\lambda}{4m}$$

as $r \to 2m$, assuming $l \ll 2m$ and using equation (4.27). After opening the box the energy is

$$\mathcal{E}'_r \sim (m_0 - \delta m_0) \frac{\lambda}{4m}.$$

The efficiency η with which the thermal radiation can be transformed into work is $[\delta m_0 - (E_r - E'_r)]/\delta m_0$, or

$$\eta = \frac{\delta m_0 - \frac{\lambda}{4m} \delta m_0}{\delta m_0}$$

$$= 1 - \frac{\lambda}{4m}.$$

Using now the value for λ, equation (4.26), this becomes

$$\eta = 1 - \frac{hc}{4mkT},$$

which has the familiar form $\eta = 1 - T_s//T$ for a sink at temperature T_s if the black hole has a temperature

$$T_h = T_s \sim \frac{hc}{4mk}.$$

That the final relation is only an approximation comes from the fact that we have used only an order of magnitude argument. We can deduce however that the temperature of the black hole in physical units satisfies

$$T_h \propto \frac{hc^3}{GMk}.$$

For a Schwarzschild black hole $\kappa = 1/(4m) = c^4/(4GM)$, where κ is the surface gravity, so in this case $T_h \propto h\kappa/ck$. (In fact this expression holds in general, not just for a spherical hole.) So in order for a black hole to satisfy the laws of thermodynamics it would need to have a temperature given by the above expression.

4.4.2 *The four laws of black hole thermodynamics*

We are now in a position to state the laws of thermodynamics for black holes.

Zeroth law

The temperature of a black hole is constant on the horizon. This follows because $T_h \propto \kappa$ and the surface gravity κ is constant on the horizon (as we showed in Section 4.1.1).

First law

The first law of black hole mechanics, equation (4.8) can now be written as

$$dm = T_h dS_h + \omega_+ dJ,$$

where $dS_h \propto dA_h$. We can therefore associate with a black hole an entropy S_h proportional to its surface area. Equivalently, $S_h \propto m^2$. With slightly more generality we can add an extra term for the charged black hole corresponding to the work done in acquiring a charge dq at an electrostatic potential Φ. The relation becomes

$$dm = T_h dS_h + \omega_+ dJ + \Phi dq.$$

This is the first law of black hole thermodynamics. Note that these arguments cannot determine the constants of proportionality in the expressions for temperature and entropy.

Second law

The direct translation of the area theorem in general relativity, that the area of a black hole cannot decrease, would be that the entropy of a black hole cannot decrease. This is not valid once quantum mechanics comes into play, because the Hawking radiation carries off mass and therefore reduces the surface area of the black hole and hence its entropy. The correct statement is a generalised form of the law, that the entropy of the universe, including that of the black hole S_h, cannot decrease, i.e. if S_{ext} is the entropy of the world excluding the black hole then $S_h + S_{ext}$ cannot decrease. For example, for an isolated radiating black hole a detailed calculation shows that, although the entropy of the black hole decreases, the increase in entropy in the external world coming from the emitted radiation exceeds the loss in entropy of the hole, so this generalised form of the second law is obeyed (Bekenstein, 1975).

Third law

The Planck statement of the third law of thermodynamics, that the entropy of any system tends to zero at zero temperature, is not valid, at least in classical black hole thermodynamics. It is violated by the extremal charged hole and the extremal Kerr hole (which have $T = 0$, $A_h \neq 0$). (The restriction to classical physics arises because extreme holes might not exist in whatever the quantum theory turns out to be.) In any case, the alternative Nernst version of the third law, that $T = 0$ cannot be reached in a finite number of steps, can be shown to hold for the temperature T_h of a black hole horizon. We have seen an example of this in our discussion of the spinning up of a black hole (Section 3.16.1). Note for $a = m$ that $T_h = 0$ (since $\kappa = 0$). So the third law corresponds in this case to the impossibility of spinning up a black hole to $a = m$.

4.5 Hawking radiation

4.5.1 *Introduction*

In this section we outline a derivation of the properties of the radiation from a Schwarzschild black hole that has been formed by collapse. The result of this calculation is that the absolute values of the temperature and entropy can be determined. We find that a black hole radiates like a blackbody with a temperature in physical units given by

$$T_h = \frac{\hbar\kappa}{2\pi ck} \tag{4.28}$$

and that its entropy is given by

$$S_h = \frac{1}{4}\frac{kc^3}{\hbar G}A_h,$$

where, as usual, $\hbar = h/2\pi$.

The details of Hawking's original paper (Hawking, 1974) can be followed with a basic knowledge of quantum field theory and complex analysis. The result has since been rederived in various ways, with minimal assumptions (see Carlip, 2009). As we hinted at earlier, there are certainly some open problems, but the consistency of the result with thermodynamics reinforces the conclusion that black holes radiate. Rather than repeat the details of one or more of the derivations in the literature, we shall try to give some insight into the main points of the physics while avoiding the full quantum field theory treatment.

The calculation treats the behaviour of an incoming quantum wave as it passes through a collapsing star and out to future (null) infinity. The major contribution to radiation from the hole comes from massless particles, since there is no threshold energy for the production of massless particle-antiparticle pairs. So, for example, for (hypothetical) massless scalar particles we should use the relativistic wave equation (4.12). Nevertheless, as long as we treat both particles and antiparticle waves explicitly, we can think of a non-relativistic many particle state (or quantum field) described by Schrödinger wave functions. Now, Schrödinger waves scatter off the gravitational field of the black hole just like classical waves, so this is not the essential ingredient introduced by quantum theory. Rather the key effect of quantum theory is felt through the properties of the quantum vacuum.

One way to understand how the vacuum has properties is to think in terms of zero point energy. This is usually introduced in the context of an isolated harmonic oscillator. Suppose however that the oscillator is coupled to an external system, for example that the oscillator consists of an electrically charged mass on a spring, which therefore interacts with the electromagnetic field. Even in its ground state the oscillating mass would radiate electromagnetic waves and so the external system would appear to drain the zero point energy from the oscillator. This would be inconsistent with quantum mechanics, in particular with the uncertainty principle. The resolution is that the electromagnetic field also has a zero point energy and just as the fluctuations in the oscillator radiate to the field, so zero point fluctuations in the field re-energise the oscillator, thereby maintaining overall balance. (This, incidentally, is why the ground state of an atom is stable.)

We can generalise this to the realisation that all fields have associated zero point energies. In fact, in relativity, we have not only the existence of positive energies but also of negative energies, or, equivalently, of positive and negative frequencies (since energy and frequency are related by a factor $\hbar$). We can see this from equation (4.14) where solutions $e^{\pm i\omega t} = e^{\pm iEt/\hbar}$ are permitted (and not only $e^{-iEt/\hbar}$ as for the non-relativistic Schrödinger equation). We can now describe the results of bona fide quantum field theoretical calculations by returning to Dirac's pre-field theory picture of the quantum vacuum. In Dirac's picture the vacuum consists of modes or states of positive and negative energy, with the negative energy states completely filled and states of positive energy empty. An anti-particle in this picture appears as the absence of a negative energy wave. (A little license is required here since Dirac's picture really applies only to fermions, so the description should not be taken too precisely.)

The crucial point will be that the quantum vacuum, in particular the negative frequency zero point modes, can be modified by the (classical) gravitational field of external matter.

4.5.2 *Casimir effect*

The best known and best established case of the distortion of the quantum vacuum by an external influence is the Casimir effect. Here the electromagnetic field interior to a system of electrical conductors, two parallel plates for example, is modified by the necessity to satisfy conducting boundary conditions on the plates. Between the conductors only certain field modes (wavelengths or, equivalently, frequencies) can exist and each contributes its $\frac{1}{2}\hbar\omega$ to the zero point energy, whereas in the exterior region all modes are present. The difference in energy density is observed as a difference in pressure, which creates forces on the conductors (pushing together parallel plates or expanding a spherical conductor). The system is static so there is no radiation of energy. But at the microscopic level one would, in principle, find real fluctuating currents in the conductors, the presence of which is required for consistency with the zero point fluctuations in the fields.

4.5.3 *Thermal vacua in accelerated frames*

The usual vacuum state of a field in the absence of gravity is defined with respect to inertial observers in flat spacetime. Instead of cluttering up this vacuum with conductors to demonstrate the effects of negative energies, we can look instead at how it appears to an accelerating observer. As we shall see, this quantum system is similar in some respects to a black hole. It is therefore a useful preparation before we treat the effects of a real gravitational field. We confine ourselves to the simplest case of a uniformly accelerated observer in one space dimension and to the vacuum states of a scalar field (rather than the electromagnetic field). Even here we shall be content to outline the discussion, leaving the interested reader to find the details in the literature (Birrell and Davies, 1982; Takagi, 1986; Unruh, 1976).

The scalar wave equation (i.e. the relativistic Schrödinger equation) in one space dimension for the field Φ in the usual inertial (or Minkowski) coordinates (t, x) is

$$-\frac{\partial^2 \Phi}{\partial t^2} + \frac{\partial^2 \Phi}{\partial x^2} = 0.$$

The separable solutions will be of the form

$$\Phi(t, x) = u_\omega(t, x) \sim \begin{cases} e^{\pm i\omega(x-t)} \\ e^{\pm i\omega(x+t)} \end{cases},$$

with $\omega > 0$. (For comparison with the perhaps more familiar form $\exp \pm i(\omega t + kx)$ note that with $c = 1$ the wavenumber k satisfies $k = \pm\omega$.) Just as in non-relativistic quantum mechanics we think of Φ as describing both a wave field and a particle state of definite momentum.

The vacuum for an inertial observer is defined by the absence of positive energy particles (or waves). However, the vacuum cannot be entirely devoid of energy, in some sense, because, even in the vacuum, fluctuations of the field must be present to maintain consistency with the uncertainty principle (corresponding exactly to the zero point energy of a harmonic oscillator). We can extend Dirac's original prescription (for fermions) and imagine that in the vacuum the positive energy states are empty but the negative energy states are populated. In fact, crudely speaking, we can think of the $\frac{1}{2}\hbar\omega$ of an oscillator to correspond to half a particle per frequency mode. In essence, we cancel this positive energy by populating the vacuum with half a negative energy particle per mode.

Positive energy solutions are eigenfunctions of the energy operator with positive eigenvalue, so

$$i\hbar\frac{\partial}{\partial t}u_\omega^{(+)} = \hbar\omega u_\omega^{(+)}$$

for positive energy solutions, and hence the time dependence of $u_\omega^{(+)}$ has the form

$$u_\omega^{(+)} \sim e^{-i\omega t},$$

with $\omega > 0$. So the allowed negative energy modes in the vacuum state are

$$\Phi \sim u_\omega^{(-)} = \frac{1}{2(\pi\omega)^{1/2}}e^{i\omega t \pm i\omega x}, \quad \omega > 0. \tag{4.29}$$

Note that we have not discussed normalisation of the wave function so we have not justified the various constants in equation (4.29), which are included for completeness. (The normalisation condition in relativity is not the same as for the non-relativistic Schrödinger equation. The reason is that we have to chose a normalisation that is preserved in time, and this depends on the evolution equation for the field. We shall ignore this technical detail here and below.) Note also that the wavenumber can be positive or negative since the negative energy waves can be incoming or outgoing. Equivalently, we can include both cases by writing

$$u_\omega^{(-)} = \frac{1}{2(\pi|\omega|)^{1/2}}e^{i|\omega|t-i\omega x}, \tag{4.30}$$

with $-\infty < \omega < +\infty$.

To find the eigenstates of the field for the uniformly accelerated observer, we transform the wave equation to the accelerated frame. This is achieved by using the Rindler coordinates (Section 2.8.3). Since we are now using

(t, x) for inertial coordinates we label the Rindler coordinates here as (η, ξ) instead of the (t, ξ) of Section 2.8.3. We have

$$t = \xi \sinh a\eta$$
$$x = \xi \cosh a\eta \tag{4.31}$$

with

$$d\tau^2 = a^2 \xi^2 d\eta^2 - d\xi^2. \tag{4.32}$$

Note that η is the proper time on the worldline $\xi = 1/a$ of the observer with proper acceleration a. The wave equation becomes

$$-\frac{1}{a^2 \xi^2} \frac{\partial^2 \Phi}{\partial \eta^2} + \frac{1}{\xi} \frac{\partial \Phi}{\partial \xi} + \frac{\partial^2 \Phi}{\partial \xi^2} = 0, \tag{4.33}$$

with $c = 1$ as usual. This has separable solutions with positive energy with respect to the time η

$$U_{\omega'}^{(+)} = \frac{1}{2(\pi |\omega'|)^{1/2}} e^{-i|\omega'|\eta} \left(\xi^{i\omega'/a} \right), \tag{4.34}$$

(with a normalisation appropriate to the relativistic wave equation) and corresponding negative energy solutions $U_{\omega'}^{(-)}$. These are the energy eigenstates according to the accelerated observer.

These solutions (4.34) form a complete orthonormal set at a fixed time η in the sense that

$$\int U_{\omega'}^{(+)} U_{\omega''}^{(+)*} \frac{d\xi}{\xi} \propto \delta(\omega' - \omega''),$$

where $\delta(x)$ is the Dirac delta function. Note the factor of ξ in the integrand is required for orthogonality.

We can prove this as follows:

$$\int U_{\omega'}^{(+)} U_{\omega''}^{(+)*} \frac{d\xi}{a\xi} = \frac{1}{2(\pi |\omega'|)^{1/2}} \frac{1}{2(\pi |\omega''|)^{1/2}} e^{-i|\omega'|\eta + i|\omega''|\eta} \int \xi^{i\omega'/a - i\omega''/a} \frac{d\xi}{a\xi}$$

$$= \frac{1}{2(\pi |\omega'|)^{1/2}} \frac{1}{2(\pi |\omega''|)^{1/2}} e^{-i|\omega'|\eta + i|\omega''|\eta}$$

$$\times \int e^{i(\omega' - \omega'')(\log \xi)/a} d(\log \xi)/a$$

$$= \frac{1}{2 |\omega'|^{1/2}} \frac{1}{|\omega''|^{1/2}} e^{-i\eta(\omega' - \omega'')} \delta(\omega' - \omega'')$$

$$= \frac{1}{2\omega'} \delta(\omega' - \omega''),$$

and we have used

$$\delta(\omega) = \frac{1}{2\pi} \int_{-\infty}^{\infty} e^{i\omega x} dx$$

with $x = (\log \xi)/a$.

Now, an inertial negative energy wave of the form (4.30) can be expressed as a superposition of the complete set of states of positive energy waves $(U_{\omega'}^{(+)})$ and negative energy waves $(U_{\omega'}^{(-)})$ for the accelerated observer as

$$u_\omega^{(-)}(t(\eta, \xi), x(\eta, \xi))$$
$$= \sum_{\omega'} \left\{ \beta_{\omega'}(\omega) U_{\omega'}^{(+)}(\eta, \xi) + \alpha_{\omega'}(\omega) U_{\omega'}^{(-)}(\eta, \xi) \right\}, \qquad (4.35)$$

where the $\beta_{\omega'}(\omega)$ and $\alpha_{\omega'}(\omega)$ are constants to be found. Strictly the sum in equation (4.35) is really an integral over continuous values of ω', but since we are interested in a mode by mode analysis we do not need to be concerned with such details.

We come now to the input from physics. Consider a quantum field which is in the mode $u_\omega^{(-)}$ of the inertial vacuum state. According to standard quantum theory, the coefficient $\beta_{\omega'}(\omega)$ on the right of (4.35) gives the amplitude for measuring a (positive frequency) particle (in the state $U_{\omega'}^{(+)}$) in the accelerated frame.

To obtain β we multiply through by $\xi^{-1} U_{\omega''}^{(+)*}$ and integrate with respect to ξ. Note that we include the factor ξ^{-1} because we want to use the orthogonality condition (4.34). The functions $U_\omega^{(-)}$ and $U_{\omega''}^{(+)*}$ can easily be shown to be orthogonal by repeating the calculation above (or simply replacing ω' by $-\omega$ and noting that $\delta(\omega' + \omega'') = 0$ if ω' and ω'' have the same sign). Thus, at any time η, for the accelerated observer in the inertial vacuum there is an amplitude for finding a particle of frequency ω' from each wave of frequency ω of

$$\beta_{\omega'}(\omega) = 2\omega' \int u_\omega^{(-)} U_{\omega'}^{(+)*} \frac{d\xi}{\xi}$$
$$= \frac{1}{2\pi} \left(\frac{\omega'}{\omega} \right)^{1/2} \int e^{i|\omega'|\eta} \xi^{-i\omega'/a} e^{i|\omega|t - i\omega x} \frac{d\xi}{\xi}, \qquad (4.36)$$

where x and t are considered to be functions of ξ at time η through (4.31) and the integral is over the 'volume' of space.) Therefore according to the accelerated observer the inertial vacuum contains an overall number of particles of frequency ω' from all of the modes of the inertial vacuum

$$n(\omega') = \int |\beta_{\omega'}(\omega)|^2 \, d\omega. \tag{4.37}$$

The integral for $\beta_\omega(\omega')$ can be evaluated in terms of gamma functions and hence the probability of detecting particles of a given frequency ω' found from equation (4.37).

However, we can simplify the calculation by making use of the fact that, since the Rindler metric is independent of time, the probability for the accelerated observer to detect particles must be independent of time. So we can evaluate this at a convenient time, which is $t = \tau = 0$. Thus we need

$$\beta_{\omega'}(\omega) = \frac{1}{2\pi} \left(\frac{\omega'}{\omega}\right)^{1/2} \int e^{-i\omega\xi} \xi^{-1-i\omega'/a} d\xi, \tag{4.38}$$

since $dx = d\xi$ on $t = 0$. To circumvent the detailed evaluation of the integral, note that, putting $z = i\omega\xi$, equation (4.38) can be written

$$\beta_{\omega'}(\omega) = \frac{1}{2\pi} \left(\frac{\omega'}{\omega}\right)^{1/2} \int e^{-z}(-iz)^{-1-i\omega'/a} \omega^{-1+i\omega'/a} i dz.$$

Writing $-i = e^{-i\pi/2}$ we get

$$\beta_{\omega'}(\omega) = \frac{1}{2\pi} \left(\frac{\omega'}{\omega}\right)^{1/2} e^{-\pi\omega'/2a} \int e^{-z} (z)^{-1-i\omega'/a} \omega^{+i\omega'/a} dz$$

$$= \frac{1}{2\pi} \left(\frac{\omega'}{\omega}\right)^{1/2} e^{-\pi\omega'/2a} I,$$

where the last equation defines I. Note the $\beta \to 0$ as $a \to 0$, consistent with the fact that the inertial observer should see just the vacuum. Thus

$$|\beta_{\omega'}(\omega)|^2 = \frac{e^{-\pi\omega'/a}}{4\pi^2} \left(\frac{\omega'}{\omega}\right) |I|^2,$$

and

$$|\beta_{-\omega'}(\omega)|^2 = -\frac{e^{\pi\omega'/a}}{4\pi^2} \left(\frac{\omega'}{\omega}\right) |I|^2 = -e^{2\pi\omega'/a} |\beta_{\omega'}(\omega)|^2.$$

Hence,

$$n(-\omega') = -e^{2\pi\omega'/a} n(\omega').$$

A function satisfying this relation is

$$n(\omega') = \frac{1}{e^{2\pi\omega'/a} - 1} \tag{4.39}$$

which is a thermal spectrum. In one space dimension the factor for the volume of phase space in the Planck spectrum, usually $4\pi k^2 dk$, is just dk $(= d\omega$ with $c = 1)$. The Planck distribution in one space dimension is therefore

$$n(\nu)d\nu = \frac{2\pi d\nu}{e^{h\nu/kT} - 1},$$

where $\omega' = 2\pi\nu$ (since ω' is in radians per unit time and ν is in cycles per unit time). Comparing this with equation (4.39) we conclude that the accelerated observer sees a thermal distribution of particles at temperature $\hbar a/(2\pi k)$, or

$$T_a = \hbar a/(2\pi kc), \tag{4.40}$$

reinstating physical units for the acceleration a.

Note that the calculation applies to an observer undergoing constant proper acceleration for all time in the inertial vacuum, and does not include any transient effects from starting and stopping the acceleration. The calculation does not tell us directly what an observer accelerated from rest can detect or what a system accelerated from rest will emit. Furthermore, while the discussion is instructive, it cannot be used directly to infer the properties of a black hole.

We can take one more message from this discussion. We ask how is it physically that the same vacuum can appear differently to two relatively accelerated observers, the freely falling Minkowski one at $x = $ constant and the accelerated one on the worldline $\xi = $ constant? The answer, in outline, is that the accelerated observer in the right Rindler wedge of the spacetime (Fig. 4.3), has no knowledge of the left-hand wedge. The quantum state of his system must therefore be described in such a way as to discard any information about the left wedge. This turns the system from a pure vacuum to a thermal state.

Problem 4.12. *Verify that* (4.34) *is a solution of the wave equation* (4.33).

4.5.4 *Hawking radiation*

In fact, one might think that it does not matter to a wave at a large distance from a spherical gravitating mass whether the mass is a star or a black

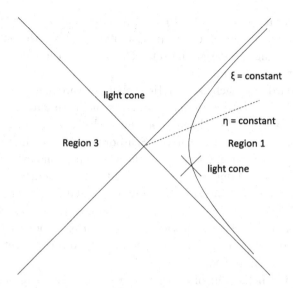

Fig. 4.3   Minkowski spacetime in Rindler coordinates (η, ξ) showing the causal region (Region 1) accessible to the observer on the constant acceleration trajectory $\xi = $ constant. This observer can neither send to nor receive signals from Region 3.

hole. A static 'star' in the inertial vacuum cannot be emitting Hawking radiation (whatever an accelerated observer might see) by the very fact that it is static. The black hole is different: its interior is not static since the metric coefficients depend on time (recall that the r coordinate is timelike inside the horizon), or, equivalently, because it possesses an event horizon (which divides the static exterior from the non-static interior). One can see clearly from the spacetime diagrams (Figs. 2.7 and 2.11 of Chapter 2) that the spacetime structure at infinity is not the same for both the star and the black hole. Physically, Hawking describes it by saying that rays traced backwards from late times pile up near $v = 2m + t_*$ (where t_* is the time at which the surface of the star crosses the horizon). This is roughly equivalent to saying that the late stages of collapse are visible for the whole of the future of an observer at infinity.

At past infinity, far from the collapsing star, we can define our usual inertial vacuum (the in-vacuum): this consists of negative energy states of the form

$$u_{\omega LM} = \frac{1}{2(\pi\omega)^{1/2}} e^{i\omega v} \quad \text{as } r_* \to \infty, \tag{4.41}$$

with $\omega > 0$. For an observer at future infinity, we shall see that the *presence of the horizon* means that the vacuum is a different state (the out-vacuum). As in the preceding section, we need to calculate how these negative energy states appear to such an observer.

The standard approach uses the Heisenberg picture in quantum mechanics, in which the states are fixed and the field operators are evolved backwards to work out the future content of the in-vacuum. Here instead we shall evolve the states as in the Schrödinger picture in introductory quantum mechanics. Rather than solve the wave equation however, we make use of the geometrical optics approximation to follow changes in the wave. This is possible because waves reaching infinity at late times are redshifted, so arise from waves that in the vicinity of the collapsing star have a high frequency, for which the geometrical optics approximation is valid. Also, spacetime near the horizon is flat (Section 2.8.3) so the geometrical optics approximation is simplified.

Consider the behaviour of a negative frequency incoming wave $e^{i\omega v}$ as it travels through the star, to emerge just at the point where the outer shell crosses the horizon, and continues to future infinity. Waves beyond $v = v_0$ go down the hole, so only waves from $v < v_0$ contribute. For simplicity we can choose the labelling such that $v_0 = 0$.

At past (null) infinity the metric is asymptotically flat in Schwarzschild coordinates. Thus we measure the frequency of a wave by a uniform spacing in Schwarzschild time t or, equivalently, by the phase difference in $\omega v = \omega(t + r_*)$. At future infinity, near $r = 2m$ we have seen that the metric is also flat. In fact, we can see that in Kruskal coordinates the metric takes the Minkowski form near $r = 2m$, so we can use our usual notions of geometrical optics near the horizon, provided that we use these coordinates and not the Schwarzschild ones. Then the frequency of a wave will be measured by constant spacing in the Kruskal U coordinate, or, equivalently by the phase $\omega'U$. In geometrical optics the phase is constant along a ray in spacetime (by definition of rays!). Changes is frequency are measured by changes of spacing between the rays. This is illustrated in Fig. 4.4 for some ingoing and outgoing waves.

Now consider a high frequency wave that comes in from infinity near $v = 0$ passing through the star as it is about to cross the horizon and emerges highly redshifted to future infinity. Note that this is illustrated in the figure by reflection at the centre of the star. This is an artefact of a spacetime diagram showing a radial coordinate and time. The incoming ray passes directly through the star, but, because the angular coordinates

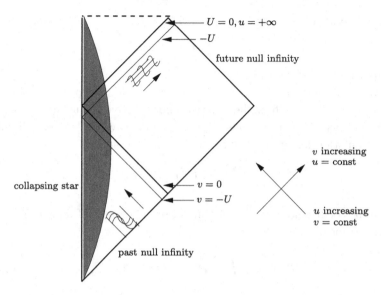

Fig. 4.4   Penrose–Carter diagram of a star collapsing to a black hole showing a ray that passes through the star and grazes the horizon at $U = 0$.

are suppressed, what we see in the diagram is the incoming ray at decreasing radial coordinate reflected into an outgoing ray at increasing radial coordinate.

We need to connect the form of the outgoing wave with that of the incoming wave. It is easy to see qualitatively what happens. The rays that fill the infinite future of an observer at a large distance from the hole are those that pile up at $v = 0$. Any ray that starts later than $v = 0$ goes straight down the hole and does not get to infinity. (We are ignoring any scattering here.) Thus a finite amount of v time in the distant past must be stretched to an infinite amount of u time in the far future. The figure shows quantitatively how this happens. The ray that grazes the horizon at a separation (in affine parameter) $-U$ arises from a ray that originates at early times far from the star at a parameter separation $v = -U$ from the last ray at $v = 0$. An incoming wave of the form $e^{\pm i\omega' v}$ (4.41) therefore becomes an outgoing wave of the form $e^{\pm i\omega' U}$.

We now express the outgoing wave $e^{i\omega' U}$, arising from the purely negative frequency incoming mode $e^{i\omega' v}$, in terms of the waves $e^{\pm i\omega u}$ that define the out-vacuum in the spacetime exterior to the hole. The two expressions are connected by the relation between Kruskal and

Schwarzschild coordinates equation (2.54):

$$e^{i\omega' U} = e^{-i\omega' \exp(-\kappa u)},$$

where κ is the surface gravity.

We must therefore expand $e^{-i\omega' \exp(-\kappa u)}$ as a superposition of waves of the form $e^{\pm i\omega u}$ to find the positive and negative frequency content at future infinity of the purely negative frequency wave $e^{i\omega' v}$ at past infinity. The calculation follows that of Section 4.5.3. We put

$$e^{-i\omega' \exp(-\kappa u)} = \sum_{\omega} \{\beta_\omega(\omega') e^{-i\omega u} + \alpha_\omega(\omega') e^{i\omega u}\},$$

where the coefficients $\alpha_\omega(\omega')$ and $\beta_\omega(\omega')$ are to be found. Taking the inverse Fourier transform, we obtain

$$\beta_\omega(\omega') \propto \int_{-\infty}^{+\infty} e^{i\omega u} e^{-i\omega' \exp(-\kappa u)} du, \tag{4.42}$$

which gives the amplitude for finding a positive frequency state $e^{-i\omega u}$ in the outgoing waves, starting from the incoming vacuum. Thus we can obtain the probability of finding particles at future infinity, in frequency (or momentum) state ω, by taking the modulus squared of the amplitude and integrating over all incoming frequencies. The result is proportional to

$$\int |\beta_\omega(\omega')|^2 \, d\omega'.$$

The number of particles emitted by the hole is also proportional to this probability.

To evaluate the integral for $\beta_\omega(\omega')$ in equation (4.42) put $x = e^{-\kappa u}$. We get

$$\beta_\omega(\omega') \propto \int_0^\infty x^{-1 - i\omega' \kappa^{-1}} e^{-i\omega x} dx.$$

This is exactly the integral we studied in equation (4.38) for the case of the accelerating observer (with x here replacing ξ). Following the same procedure we therefore obtain a flux of particles (number per unit time in the frequency interval $d\omega$) with a spectrum proportional to

$$\frac{d\omega}{e^{2\pi\omega/\kappa} - 1}, \tag{4.43}$$

which is that of a blackbody with temperature proportional to κ.

Note that we have approximated the spacetime as flat at large distances and close to the horizon. More accurately we should use the Regge–Wheeler equation (4.17) or the Teukolsky equation (4.22) for the propagation of waves in the Schwarzschild or Kerr geometries. If the potentials in these equations are taken properly into account then there is an overall factor in equation (4.43) arising from the scattering of the waves off of the potentials. Since the potentials are different for particles of different spin the rate of emission of particle species from the black hole will differ, but the spectrum (i.e. the distribution of particles over different frequencies) will remain the same. These transmission factors are ignored below.

The outline of the calculation we have given here is clearly far from rigorous. The underlying calculation has now been picked over many times and redone in many different ways which leaves no doubt about the conclusion. We shall look at some outstanding issues in Section 4.7.

Problem 4.13. *Near $r = 2m$ show that*

$$\frac{r_*}{2m} - 1 \approx \log_e \left(\frac{r}{2m} - 1 \right)$$

and that

$$1 - \frac{2m}{r} \approx \frac{r}{2m} - 1.$$

Deduce that

$$\kappa^2 \xi^2 \approx \exp \left(\frac{r_*}{2m} - 1 \right)$$

and hence the Kruskal coordinate $U = -e^{-u/4m}$ satisfies

$$U \propto \xi e^{-\kappa t}.$$

By comparing with the transformation from Rindler coordinates (t, ξ) to flat space (Minkowski) null coordinates $u = T - X$ (and $v = T + X$) deduce that the metric near $r = 2m$ has approximately the Minkowski form in Kruskal coordinates. Confirm this directly (and more simply!) from the Kruskal form of the metric near $r = 2m$.

4.5.5 *The temperature of a general black hole*

In this section we give a shortcut to calculating the temperature of a black hole. Let us return to the Rindler metric (equation (4.32)) but with a Euclidean signature:

$$ds^2 = \xi^2 d\theta^2 + d\xi^2 \tag{4.44}$$

where $\theta = iat$ and $ds^2 = -d\tau^2$. We have written our new 'time' coordinate as an angle θ because we are now going to take it to be periodic with period 2π. The Rindler metric is thereby transformed into the metric on the plane in polar coordinates. If θ has period 2π then it must have period $2\pi/a$. It can be shown that what is the Minkowski vacuum for an inertial observer will correspond to a thermal state for the Rindler observer on the trajectory $\xi = 1/a$ with an inverse temperature $\beta = 1/kT$ equal to the periodicity in time t, $\beta = 2\pi/a$. Thus the uniformly accelerated observer sees the Minkowski vacuum to have a temperature $T_a = a/2\pi k$, which, reinstating the constants c and $\hbar$, is just what we found in equation (4.40). We can use this approach to obtain the temperature of a general black hole because the near horizon metric can always be approximated by the Rindler metric. For example, for the Schwarzschild case, from (2.42), the near-horizon metric (with the angular coordinates suppressed) takes the form (4.44) if

$$it = \kappa\theta$$

where κ is the surface gravity. Thus the temperature of a Schwarzschild black hole in equilibrium with thermal radiation is $T = 2\pi/\kappa$. This is therefore also the temperature at which the black hole will radiate, in agreement with equation (4.28).

4.6 Properties of radiating black holes

4.6.1 *Entropy and temperature*

The quantum mechanical calculation fixes the constants in the expression for the temperature and entropy of a black hole. Restricting ourselves to the Schwarzschild case, in physical units these are

$$T_h = \frac{\hbar c^3}{8\pi k G M} \sim \frac{10^{23}}{M} \text{ K}$$

or about $10^{-7}(M_\odot/M)$ K and

$$S_h = \frac{1}{4}\frac{kc^3}{\hbar G}A_h \sim 10^{77}\left(\frac{M}{M_\odot}\right)^2 k \quad \text{JK}^{-1}.$$

It is of interest to estimate the thermal entropy of a star before collapse to a black hole. If the star contains N particles the entropy is of order $S \sim Nk$. For a solar mass star ($M_\odot \approx 10^{30}$ kg) this is $S \sim 10^{57}k$. The difference between the entropy of the star and the black hole it forms is therefore enormous. What is the source of this entropy? We shall return to this issue in Section 4.7.

We can define 'natural' units of mass (or energy), length and time in terms of the fundamental constants G, $\hbar$ and c, respectively the Planck mass (or energy), length and time, as

$$m_P = \left(\frac{\hbar c}{G}\right)^{1/2}, \quad E_P = m_P c^2, \quad l_P = \left(\frac{\hbar G}{c^3}\right)^{1/2}, \quad t_P = \left(\frac{\hbar G}{c^5}\right)^{1/2}.$$

In terms of these quantities the horizon temperature is

$$\frac{kT_h}{m_p c^2} = \frac{1}{8\pi}\left(\frac{m_P}{M}\right)$$

and the entropy is

$$S_h/k = \frac{1}{4}\left(\frac{A_h}{l_P^2}\right).$$

Thus $S_h/k = \frac{1}{4}\times$ the area of the black hole in units of the square of the Planck length $(\hbar G/c^3)^{1/2}$.

Problem 4.14. *Show that the entropy of a Planck mass spherical black hole is $4\pi k$.*

4.6.2 *Radiating black holes*

The rate of energy loss by a Schwarzschild black hole is obtained from Stephan's law as follows:

$$L_h = A_h \sigma T^4 = 16\pi \left(\frac{GM}{c^2}\right)^2 \sigma \left(\frac{\hbar c^3}{8\pi GMk}\right)^4$$

where we have considered electromagnetic radiation only as this will make a major contribution to the emitted radiation. Inserting numerical values we get for the luminosity of the hole

$$L_h = 3.6 \times 10^{32} M^{-2}$$

$$= 9.0 \times 10^{-29}\left(\frac{M_\odot}{M}\right)^{-2} \text{W}.$$

Since this must come from a decrease in the mass of the hole, we have

$$\frac{dM}{dt} = -\frac{3.6 \times 10^{32}}{c^2 M^2}.$$

integrating gives the lifetime of a black hole of mass M as

$$t_h \sim 8.33 \times 10^{19} \left(\frac{M_i}{10^{12}\ \text{kg}}\right)^3 \text{s}$$

where M_i is the initial mass of the hole. So a black hole of mass between 10^{11} kg and 10^{12} kg would be exploding today.

Problem 4.15. *What is the lifetime of a solar mass black hole?*

Problem 4.16. *Show that a black hole of mass 1.6×10^{12} kg has a lifetime of about 10^{10} years (a Hubble time). What is the lifetime of a black hole that is just hot enough to be emitting gamma rays?*

Problem 4.17. *What is the temperature of a Planck mass black hole? What is the entropy of a black hole of area 1 m^2.*

Problem 4.18. *The specific heat of a body of mass M is defined as $d(Mc^2)/dT$. Show that the specific heat of a Schwarzschild black hole of mass M is*

$$C = -\frac{8\pi kG}{\hbar c}M^2.$$

Note the negative sign: a black hole heats up as it radiates as is often the behaviour of gravitating systems. (The paradox is resolved by noting that we are leaving out of account the gravitational energy.)

Problem 4.19. *Show using the method of Section 2.8 that the surface gravity of a Reissner–Nordström black hole is $\kappa = (mr_+ - q^2)/2r_+^3$ and hence that the temperature of an extreme Reissner–Nordström black hole is zero.*

4.6.3 *Black hole in a box*

Consider first a black hole in thermal equilibrium with a blackbody radiation field held at a fixed temperature, for example in a large thermal reservoir. The radiation will be affected by the gravitational field of the hole because corresponding modes of the field are relatively redshifted at different radii. The parameter T in the Planck distribution is therefore not the local temperature, but the temperature at infinity. The standard form of the Planck distribution takes this into account without modification.

Suppose that a fluctuation causes the emission of the black hole to increase by a small amount. Its mass will fall and its temperature will rise. But accompanying the fall in mass will be a decrease in the area of the black hole, so it will absorb less energy from the radiation field, the temperature of which we have supposed not to vary. Thus the departure from equilibrium grows. Likewise a fluctuation that causes the emission of the black hole to decrease by a small amount will shift the black hole away from equilibrium in the opposite sense. In this situation, therefore, thermal equilibrium is unstable. This outcome is a consequence of the negative specific heat of the black hole (Hawking, 1976).

Nevertheless there are other circumstances under which a black hole can come into thermal equilibrium with radiation in a closed system. Consider a body radiating into an isolated enclosure initially containing no radiation. As the body radiates it heats up the enclosure. A normal hot body will cool as it radiates so the body and the enclosure will eventually come into thermal equilibrium at a common temperature. This follows no matter how large the enclosure: the larger the enclosure the lower the equilibrium temperature. On the other hand, the temperature of a black hole increases as it radiates into the enclosure. So can the temperature of the enclosure catch up that of the black hole to achieve a state of thermal equilibrium?

The answer is that thermal equilibrium can be achieved only if the volume of the enclosure is less than a certain maximum size. For larger volumes the black hole evaporates away before an equilibrium can be reached. The following demonstration is based on the treatment of Custodio and Horvath (2003).

For a blackbody of initial mass M_i in an enclosure of volume V containing blackbody radiation with energy density $\rho(t)$, conservation of energy implies

$$M_i c^2 = M(t)c^2 + \rho(t)V,$$

where $M(t)$ its mass of the black hole at time t. Thus, using $\rho = aT^4$, the temperature of the radiation at time t is given by

$$T(t) = \left[\frac{(M_i - M(t))c^2}{aV} \right]^{1/4}.$$

The temperature of the black hole at time t, from Section 4.6.1, is $T = B/M(t)$ where $B = 1.12 \times 10^{22}$ kg K. So the condition for thermal equilibrium is

$$\left[\frac{(M_i - M_{eq})c^2}{aV}\right]^{1/4} = \frac{B}{M_{eq}},$$

which can be rearranged to give

$$\left(1 - \frac{M_{eq}}{M_i}\right)\left(\frac{M_{eq}}{M_i}\right)^4 = \frac{aVB^4}{c^2 M_i^5}. \tag{4.45}$$

The left-hand side of this equation has a maximum value of 0.082. This imposes a maximum value on the right-hand side and hence, for a given initial black hole mass, on the volume V of the box. For a box larger than the maximum size the black hole evaporates away before thermal equilibrium is reached.

Problem 4.20. *Confirm that the maximum value of the left-hand side of equation (4.45) is 0.082. What is the maximum volume of a box in which a Planck mass black hole can reach equilibrium?*

4.7 Entropy and microstates

In this section we shall consider a number of problems that arise from the thermodynamics of black holes. These are the entropy problem and the information problem or 'paradox'.

4.7.1 *The entropy problem*

In statistical physics the entropy, S, has a meaning in terms of the number, W, of microstates corresponding to a given macrostate through the Boltzmann relation

$$S = k \log W. \tag{4.46}$$

We have seen that associating the entropy of a black hole with its area in Planck units leads to a large entropy. But a black hole seemingly has no microstates. A black hole is uniquely determined by a few macroscopic parameters: in the case of a Schwarzschild black hole just its mass, and for a Kerr black hole its mass and angular momentum. To be clear, compare this with a volume of gas: a given (macroscopic) energy certainly does not define the microstate uniquely, but corresponds to a large number of internal microstates. According to the Boltzmann relation (4.46) the entropy of a black hole should be zero. Turning this round, we could imagine that finding the large entropy of a gas from thermodynamics would lead us to search for the internal degrees of freedom of the gas, which would lead to the atomic theory. So, we might expect that discovering the microstates that account for the entropy of a black hole would lead to a correct theory of gravity. In the early days of black hole physics, following Hawking's original discovery of black hole radiation, there were high hopes that black holes would play for unified theories of gravity the role that radiation physics had played in Planck's discovery of the quantum theory. Indeed, the Hawking formula for the black hole temperature involves gravity (G), relativity (c) and quantum theory (h). However, this does not necessarily mean that the process involves a quantum theory of gravity itself. For example, the expression for maximum mass of a white dwarf star, in which the pressure support against gravity is provided by a Fermi sea of electrons with relativisitic energies, involves G, h and c but not quantum gravity. It would seem that around a black hole with a mass much greater than the Planck mass classical gravity should be sufficient. Recent calculations of the black hole entropy from string theory suggest that this conclusion might be wrong and that the quantum nature of gravity, in the guise of string theory, might be needed to resolve the outstanding issues of black hole evaporation discussed in the following sections. In as far as it is possible at the level of this book, we shall attempt to give some idea of how this comes about in Section 5.8.

4.7.2 *The information problem*

The second problem also has to do with entropy but is more subtle. The easiest way to approach it is to consider the radiation from a burning lump of coal and to compare this with the emission by a black hole.

Before there is any emission of radiation from the coal we can describe the state of the system as the product of the state of the coal and the state of the environment, since the two are independent. Let us now look at the radiation from the coal one photon at a time. To avoid unnecessary complication we imagine that burning coal comprises emission from excited 'coal atoms'. We start with the first excited atom that is about to emit a photon. This is described by a quantum state $\Psi = \psi'_{a1}\Psi'$, say, where ψ'_{a1} describes the excited state of the first coal atom and Ψ' describes the state of the rest of the coal. The atom now emits a photon. For the sake of simplicity, let us assume that this photon can be in one of two quantum states, left $(-)$ or right $(+)$ circularly polarised, say. If the photon is circularly polarised in one direction, the atom will have an opposite spin so as to conserve angular momentum in the emission process. Then the state of the system after emission is proportional to

$$\Psi = [\psi_{a1}(+)\psi_{p1}(-) + \psi_{a1}(-)\psi_{p1}(+)]\Psi'$$

where $\psi_{p1}(\pm)$ describes the state of the first emitted photon. The system is still in a pure quantum state (described by a wave function), but the state is not a product of wave functions of the individual parts. We say that the photon is entangled with the coal (or specifically with atom 1 of the coal). If we were to measure the state of the emitted photon and find, say, that it is $\psi_{p1}(+)$, then the wave function of the system would collapse to

$$\Psi = \psi_{a1}(-)\Psi' \times \psi_{p1}(+)$$

which is again a product state in which we can describe the radiation and the coal separately.

Assuming however that we do not make a measurement on the first photon, after the emission of the second photon we have:

$$\Psi = [\psi_{a1}(+)\psi_{p1}(-) + \psi_{a1}(-)\psi_{p1}(+)]$$
$$\times [\psi_{a2}(+)\psi_{p2}(-) + \psi_{a2}(-)\psi_{p2}(+)]\Psi''.$$

When we measure some observable property $\hat{O}$ of the radiation, we are not normally interested in the detailed description of the state of $N \sim 10^{23}$ coal atoms that might go along with this. So we throw this information away by averaging over the states of the coal. For two atoms:

$$_a\langle\Psi^*|\hat{O}|\Psi\rangle_a = |\psi_{a1}(+)|^2|\psi_{a2}(+)|^2\langle\psi_{p1}^*(-)\psi_{p2}^*(-)|\hat{O}|\psi_{p1}(-)\psi_{p2}(-)\rangle$$
$$+ |\psi_{a1}(+)|^2|\psi_{a2}(-)|^2\langle\psi_{p1}^*(-)\psi_{p2}^*(+)|\hat{O}|\psi_{p1}(-)\psi_{p2}(+)\rangle + \cdots,$$

and in general

$$_a\langle\Psi^*|\hat{O}|\Psi\rangle_a = (|\psi_{a1}(+)|^2|\psi_{a2}(+)|^2 \ldots |\psi_{aN}(+)|^2)$$
$$\times \langle\psi_{p1}(-)^*\psi_{p2}(-)^* \ldots \psi_{pN}(-)^*|\hat{O}|\psi_{p1}(-)\psi_{p2}(-)\cdots$$
$$\psi_{pN}(-)\rangle + \cdots$$

where the sum on the right is over all possible products $\psi_{p1}(\pm)\psi_{p2}(\pm)\cdots$ $\psi_{pN}(\pm)$.

In general (except in the case that the photon states happen to be eigenstates of the observable $\hat{O}$) this expression cannot be rewritten in the form of the expectation value of a wave function for the photon. When we cannot describe a quantum system by a wave function, we say that the system is not in a pure state but is in a mixed state. (The description of a mixed state uses a density matrix instead of a wave function.) Note that the description in terms of a mixed state is introduced here for convenience: we have chosen to discard all the correlations between the radiation and the coal atoms, because there is no practical observation in which we would need all of the information about the coal atoms. However, the system of photons and coal, as a whole, is in principle in a pure state.

To get a bit closer to an analogy for black hole evaporation, imagine that photon 2 interacts with atom 1, for example by scattering. Photon 2 is now correlated (entangled) with atom 1 and hence with photon 1. In this way we can imagine that the radiated photons carry all the entanglement information. But we would still never be in a position where we would want to, or be able to describe the system in this much detail. So we would still want to *describe* the radiation by a density matrix, but that would be our choice: the system would still be in a pure state.

Now compare the 'evaporation' of the coal with the evaporation of a black hole. The first photon (or other particle) is emitted to infinity, and a companion negative energy particle falls into the hole. These two particles are entangled. For example, if one photon is right circularly polarised the other must be left polarised to conserve angular momentum. Now the second photon is emitted; it cannot be entangled with the first photon that went into the hole because the two are causally unrelated. (No signal can pass between them as is clear in the Kruskal diagram, Fig. 4.5.) If we keep track of all the photons we can describe the system as a pure state; and if we average over the photons in the hole we describe the radiation emitted to infinity in a mixed state.

But what is the situation when the hole has evaporated completely? Now there are no photons inside the event horizon for those outside to be entangled with, but the radiation at infinity is still described by a mixed state. This contravenes the basic tenets of quantum physics, that evolution is unitary or, more simply that probabilities are conserved. In standard quantum theory it is not possible for a pure state to evolve to a mixed one because this would entail a real loss of information. This contradiction between black hole physics and quantum theory is called the *information paradox*.

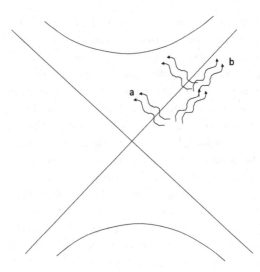

Fig. 4.5   Photons falling into the hole at early times labelled 'a' are causally disconnected from those emitted to infinity at later times at 'b'.

Various proposals have been made to resolve the paradox. Since the final stages of evaporation of a black hole involve distances, times and masses on the Planck scale, we expect that quantum gravity will be involved at this point, which at the current state of ignorance means we cannot predict the outcome. Thus, it is conceivable that a black hole does not evaporate completely but leaves a Planck mass remnant. It is however difficult to see how to endow such a remnant with enough degrees of freedom to contain the whole entropy of an arbitrary initial state.

If we now agree that there is no remnant and that the evolution must be unitary we get a different problem called a firewall. In order to achieve a pure state for the radiation from an evaporating black hole, the radiation emitted at late times must be entangled with radiation emitted at earlier times, as it was for the coal. Let us work out what an observer falling freely across the horizon would see in this case. Recall from the end of Section 4.5.3 that for the uniformly accelerated observer the interior Rindler wedge is inaccessible so her vacuum is described in terms only of external (Rindler) quanta. The Minkowski vacuum on the other hand is described in terms of quanta that are entangled between the Rindler wedges. In the Minkowski vacuum the uniformly accelerated observer sees a thermal spectrum of radiation because the correlations between the wave functions in the right and left Rindler wedges are broken by averaging over the causally disconnected wedge. The freely falling (inertial) observer in the Minkowski vacuum obviously sees nothing.

The situation near the horizon of a black hole is analogous, because close to the horizon the metric is approximately Rindler for the stationary (uniformly accelerating) observer at fixed radial coordinate and Minkowski for the freely infalling observer. The observer falling into a black hole will see nothing – the Minkowski vacuum – if the exterior quanta are (maximally) entangled with the interior quanta. The stationary observer will see a thermal spectrum because for her the interior is inaccessible, so her vacuum is described in terms only of external quanta.

Now, it turns out that there is a theorem in quantum mechanics that a system cannot be (maximally) entangled with more than one other system. (This is, in effect, equivalent to what is known as the 'no-cloning theorem' in quantum information theory.) But if the near horizon quanta are un-entangled with the quanta interior to the hole then it is the *infalling* observer who will see a thermal spectrum; this is the firewall. The result appears to contradict the equivalence principle, which is a basic tenet of general relativity. The equivalence principle says that locally an observer cannot distinguish between free fall in a gravitational field and in the absence of gravity; clearly a firewall in one case and not in the other would be a pretty big distinction.

As we shall see in Chapter 6, quantum gravity, in the guise of string theory, may come to the rescue for all of these problems.

Chapter 5

Wormholes and Time Travel

5.1 Introduction

A spacetime metric, obtained from Einstein's equations, is a purely local quantity, which does not determine the global topology of spacetime. To illustrate the meaning of this statement take the analogy of a flat sheet of paper of zero thickness rolled into a cylinder. The metric on the flat sheet and on the cylinder are the same flat Euclidean metric, as is clear from the fact that distances between neighbouring points are not altered by rolling up the sheet without stretching or tearing it. But global relations *are* changed: points near opposite edges on the flat sheet, that were separated by the length of the sheet, are close together on the cylinder. The cylinder provides a short cut between certain points that were previously far apart. We can even (in principle) roll the cylinder up in its other dimension to form a torus, which is also locally indistinguishable from flat space as far as its metric properties are concerned. To avoid confusion, recall that a two dimensional surface may appear curved from the point of view of the three dimensional space in which it is embedded although measurements confined to the two-dimensional space, for example on the sum of the angles of a triangle, will show it to be intrinsically flat. Both the cylinder and torus are intrinsically flat although they appear to be curved from the point of view of the embedding space.

We can imagine rolling up (two-dimensional) Minkowski spacetime in a similar way. We can also generalise the construction to four dimensions. Consider two inertial observers at relative rest. We identify points along these worldlines at the same times so that the intervening space is wrapped up into a cylinder. A traveller arriving at event 1 in Fig. 5.1 finds himself at event 2, so there are now two paths between 1 and 2: the identification has provided the additional short cut between the two locations. We call this shortcut a wormhole. This is not a time machine or a means of faster than light travel. The outcome arises from the topology of a cylindrical

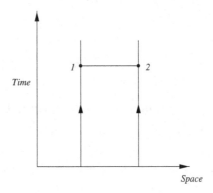

Fig. 5.1 Points at the same time on the two world-lines are identified, that is, deemed to be the same event, thereby creating a shortcut in zero time between the world-lines.

spacetime: the distance between the two points the short way round is zero.

5.2 Wormholes

The problem with the model of the previous section is that we do not have a practical means of identifying points in Minkowski spacetime to create the initial shortcut. Can wormhole metrics be created from gravitating matter and made to behave in the same way?

Our first thought might be to take the complete Schwarzschild spacetime (defined by the Kruskal metric) and roll it up such that the region IV in Fig. 2.6 of Chapter 2 is identified as part of our own universe and not a separate one. We could then see if we could travel through the black hole to a different region of spacetime, possibly in such a way that enables us to arrive at a distant destination faster than a light ray taking the conventional route. This proves to be impossible. Figure 5.2 shows why. The embedding diagram of Fig. 2.9 in Chapter 2 does not take into account the change in the Schwarzschild metric with time interior to the event horizon. When we attempt to cross from region I to region IV, it turns out that the throat pinches off just as we arrive to go through it. Another way of saying this is that however we travel through the hole there is no way to avoid the singularity.

The Kerr solution (Fig. 3.6 of Chapter 3) seems to offer additional possibilities to avoid the singularity, which is here timelike (Section 3.18). But in

Fig. 5.2 The interior of the Schwarzschild solution is dynamic: this is represented in the embedding diagram as a pinching off of the throat.

this case the spacetime structure is unstable to the small perturbations which would arise if we were to try to move any gravitating matter or energy between asymptotic regions. The instability alters the spacetime structure and would appear to lead to a spacelike singularity (Frolov and Novikov, 1998).

Our conclusion is that black holes will not enable us to construct short cuts across the universe. However, it turns out that there are other ways in which we might achieve this through what are called traversible wormholes.

5.3 Traversible wormholes

In this section we look at metrics for traversible wormholes. We shall approach the issue just as in previous chapters by giving the metric of a simple example of a wormhole spacetime and investigating its properties. However, in contrast to our approach in Chapters 2 and 3, we shall not insist that the spacetimes are empty of matter, since this would rule out any interesting wormhole solutions. We shall postulate a metric and use Einstein's equations to tell us the properties of the gravitating material needed to produce it, in contrast to the usual procedure by which the metric is obtained by solving Einstein's equations for a given matter distribution.

A general metric for a wormhole takes the form

$$d\tau^2 = e^{2\Phi(r)}dt^2 - \frac{dr^2}{1 - b(r)/r} - r^2\left(d\theta^2 + \sin^2\theta d\phi^2\right) \qquad (5.1)$$

where Φ and b are functions of r. The metric is spherically symmetric and static. The minimum value of r occurs where $b(r) = r$, say at $r_{\min} = b_0$. This defines the throat of the wormhole. At infinity $\Phi(r) = 0$; otherwise it is negative and finite, so a stationary particle always has a timelike worldline: there is no horizon. The proper distance from the throat is l given by

$$dl = \frac{dr}{(1 - b(r)/r)^{\frac{1}{2}}}. \tag{5.2}$$

A simple special case of a wormhole metric is obtained by putting $\Phi = 0$, $b = b_0^2/r$, with $r^2 = l^2 + b_0^2$,

$$d\tau^2 = dt^2 - dl^2 - \left(b_0^2 + l^2\right)\left(d\theta^2 + \sin^2\theta d\phi^2\right) \tag{5.3}$$

which is the case we shall discuss. The embedding diagram for this metric is given in Fig. 5.3. In contrast to the Schwarzschild metric, which is time-dependent inside the horizon, the wormhole metric is static everywhere. Thus the embedding diagram gives the properties at all times. An observer can travel through the throat and emerge at the other side in a finite time (problem 5.2). There is therefore no singularity in the way. This is why the spacetime is referred to as a traversible wormhole. An alternative way to demonstrate the absence of a singularity is to calculate the curvature. This also turns out to be finite everywhere.

Notice that for outwardly propagating radial light rays $dt = dl$. Thus a radial light ray can always escape to infinity, so this is another way of seeing that the spacetime has no horizon. There are two regions, $l \to \pm\infty$, where the curvature tends to zero at large distances from the wormhole.

Fig. 5.3   The embedding diagram for the special wormhole metric (5.3).

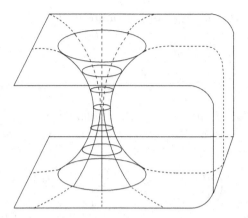

Fig. 5.4 If the mouths of a wormhole connect regions of the same spacetime the wormhole provides a short-cut.

These regions are joined by the throat of the wormhole. There is no restriction on travel between the two regions.

How could this provide a short cut across space? To see in principle how it could be done recall that we are free to change the topology without affecting the metric. Figure 5.3 shows how we can (in theory) join spacetime to create a shorter route through a wormhole.

In the wormhole metric (5.1) a particle with energy per unit mass E has a speed $v = (1 - e^{2\Phi}/E^2)$ relative to a stationary observer. (See problem 5.2.) In the metric (5.3), we have $e^{2\Phi} = 1$, so v depends only on the energy of the particle and not on its position. This tells us that the wormhole does not accelerate (i.e. attract) particles in its environment. For the more general metric (5.1) with Φ a function of r, if $E = 1$ then $v = 0$ at infinity, but we have $v > 0$ at finite values of r. So, in the field of this metric, particles are attracted towards the wormhole.

Problem 5.1. *Show that the embedding diagram of Fig. 5.3 is given by $z = b_0 \cosh^{-1}(r/b_0)$. Calculate the circumference of the throat of the wormhole.*

Problem 5.2. *In the wormhole metric (5.1) show that relative to a stationary observer the velocity of a body with energy per unit mass E in radial free fall is given by*

$$v = \left(1 - \frac{e^{2\Phi}}{E^2}\right)^{\frac{1}{2}}. \tag{5.4}$$

(*This illustrates that as long as* Φ *is finite then there is no region where* $v = c$ *so there is no horizon.*) *Show that the coordinate time to traverse a wormhole from* $-L$ *to* $+L$ *is*

$$\int_{-L}^{+L} \frac{dl}{e^{\Phi}v}. \tag{5.5}$$

Problem 5.3. *Show that the radial coordinate of a light ray of energy at infinity* E_{ph} *and angular momentum* L_{ph} *in the simple wormhole metric* (5.3) *satisfies*

$$\left(\frac{dr}{d\lambda}\right)^2 = \left[E_{ph}^2 - \frac{L_{ph}^2}{r^2}\right]\left(1 - \frac{b_0^2}{r^2}\right).$$

By looking at the zeros of the right-hand side, show that light with impact parameter $L/E = b_0$ *falls into the wormhole, and hence that the capture cross-section for a light beam is* πb_0^2.

Problem 5.4. *Show that the proper acceleration* a *felt by an observer stationary in the wormhole metric* (5.1) *at radial coordinate* r *is given by*

$$a = \pm\left(1 - \frac{b}{r}\right)^{\frac{1}{2}}\Phi'. \tag{5.6}$$

(*Hint: Obtain* $a^{\mu} = Du^{\mu}/d\tau$ *and use* $a^{\mu}a_{\mu} = -a^2$ (*Section 1.3.9*).)

This shows that if $\Phi \neq$ constant, rocket motors are required to remain at rest; otherwise the acceleration is zero.

5.4 Creating a wormhole

Can the wormhole solutions of Section 5.3 be created from gravitating matter and constructed to provide shortcuts across spacetime? A problem arises when we ask what sort of matter we need in order to create the curvature of the wormhole metric. Substituting the metric coefficients from (5.3) into Einstein's equations (1.14) shows us the matter content corresponding to the wormhole metric. This is quite a lengthy calculation the result of which is (Morris and Thorne, 1988)

$$T^{00} = T^{11} = -\frac{1}{8\pi G}\frac{b_0^2}{(b_0^2 + l^2)^2} \tag{5.7}$$

with the tangential components of equal magnitude but opposite sign. The quantity T^{00} is the density of mass-energy and T^{11} is the radial pressure. Note that the energy density and radial pressure are both negative!

Negative energy densities in the local rest frame of the matter are not possible in classical physics. More generally, classical physics forbids energy densities which are positive in the local rest frame but negative to rapidly moving observers. This is called the 'weak energy condition' which we deal with in the next section. In quantum mechanics negative energy densities are possible and we shall look at these in Section 5.6.

5.5 Weak energy condition

Let $T^{\mu\nu}$ be the energy-momentum tensor of a fluid and let v^μ be the 4-velocity of an observer. When the observer is *stationary* in the rest frame of the fluid, $v_\mu = v^\mu = (1, 0, 0, 0)$ the scalar product

$$T^{\mu\nu} v_\mu v_\nu = T^{00} = \rho, \tag{5.8}$$

where ρ is the mass-energy density of the fluid in its rest frame. By the tensor transformation law, if the observer has a velocity v^μ relative to the fluid, she will measure an energy density

$$T^{0'0'} = v^\mu v^\nu T_{\mu\nu}.$$

Therefore, if all observers measure a positive energy density, a condition we would expect all normal matter to obey, we must have

$$T^{\mu\nu} v_\mu v_\nu \geq 0.$$

This is called the *weak* energy condition. Let us specialise to the case of a perfect fluid, for which, by definition, the energy-momentum tensor is

$$T^{\mu\nu} = (p + \rho)u^\mu u^\nu - pg^{\mu\nu}.$$

In a local inertial frame $v^\mu = (\gamma, \gamma v, 0, 0)$ and $v_\mu = (\gamma, -\gamma v, 0, 0)$, where $\gamma = (1 - v^2)^{-1/2}$. Then

$$T^{\mu\nu} v_\mu v_\nu = \gamma^2(\rho + v^2 p).$$

Thus, if the weak energy is to be obeyed for all speeds $v \leq 1$ (where $v = 1$ is the speed of light in our units), it follows that, for a perfect fluid,

$$p + \rho \geq 0.$$

In the wormhole spacetimes the weak energy condition is violated, so some observers will measure a negative energy density. We call material with a negative energy density 'exotic matter'. The possibility of creating wormholes therefore comes down to the viability of creating stable exotic matter.

Problem 5.5. *The stress-energy tensor of a perfect fluid is given by*

$$T^{\mu\nu} = (\rho + p)\, u^{\mu} u^{\nu} - p g^{\mu\nu}. \tag{5.9}$$

Verify that

$$T^{\mu\nu} v_{\mu} v_{\nu} = \gamma^{2}(\rho + v^{2}p) \tag{5.10}$$

where v^{μ} is the 4-velocity of an arbitrary observer moving with speed v in the x-direction with respect to the rest frame of the fluid and $\gamma = (1 - v^{2})^{-1/2}$.

Problem 5.6. *Vacuum energy has the equation of state $p = -\rho$, with ρ positive. Show that vacuum energy does not violate the weak energy condition. Show also that the vacuum energy has the same value in all inertial frames, as it should to be consistent with the principle of relativity.*

The equation of state of cosmological dark energy has the form $p = \omega\rho$. Observations imply $\omega \approx -1$. Show that if $\omega < -1$ then the weak energy condition is violated. Such a universe would end in a 'big rip': structures on all scales down to the smallest would be torn apart.

5.5.1 *Other energy conditions*

For completeness we can note here other conditions on the energy momentum tensor that are commonly used in theorems.

- The dominant energy condition: This is the requirement that energy cannot flow faster than the speed of light locally. In general, the condition is that $-T^{\mu}_{\nu} v^{\nu}$ is timelike or null if v^{μ} is timelike or null. For a perfect fluid the requirement reduces to $\rho \geq |p|$.
- The strong energy condition: In general, for an observer with (timelike) velocity v^{μ}, the condition is $T^{\mu\nu} v_{\mu} v_{\nu} \geq \frac{1}{2} T v^{2}$ where $T = T^{\mu}_{\mu}$. For a perfect fluid the strong energy condition is $\rho + p \geq 0$ and $\rho + 3p \geq 0$. Note that the weak energy condition can be violated even if the strong energy condition is satisfied, but not if the dominant energy condition holds.
- Null energy condition: If n^{μ} is a future pointing null vector (i.e. $g_{\mu\nu} n^{\mu} n^{\nu} = 0$) then $T_{\mu\nu} n^{\mu} n^{\nu} \geq 0$.

5.6 Exotic matter

Consider two ordinary flat metallic plates held parallel and separated by some distance a. These form a capacitor. There exists an attractive force between the plates which increases with decreasing separation. One can think of this as a force between the fluctuating currents in the metallic boundaries. These do not cancel because they are correlated, the currents in one plate inducing those in the other and vice versa. Since the force is attractive the energy density between the plates must be less than that of the exterior vacuum, that is, it must be negative:

$$\rho = -\frac{\pi^2 \hbar}{720 a^4}. \tag{5.11}$$

This is called the Casimir energy and has been extensively verified experimentally (Bordag *et al.*, 2001). The corresponding energy density of the electromagnetic field on the horizon of a black hole is (Jensen *et al.*, 1988)

$$\rho = -\frac{19 \hbar}{7680 \pi^2 M^4}. \tag{5.12}$$

The amount of negative energy involved in a laboratory-scale capacitor is minuscule compared to that required for a macroscopic wormhole. It is not clear if large amounts of negative energy, sufficient to form a macroscopic wormhole, can be created. However, there is at present no known reason why, in principle, this should not be possible. We shall assume in what follows that wormholes can be made by a sufficiently advanced technology. This assumption forces us to face a fundamental issue: if a wormhole can be constructed then so too can a time machine.

5.7 Time machines

Let us assume that the topology of flat spacetime has been altered to contain a wormhole. In this section we explain how such a wormhole can be turned into a time machine (Visser, 1995).

Consider two worldlines A and B at relative rest. These will approximate the two mouths of a traversible wormhole. As in Section 5.1 to make the wormhole, we identify points on A and B that correspond to equal times in their rest frame O. The short cut through the wormhole takes zero time in this approximation: a traveller at an event on A has the choice of travelling to B through the external space or, taking a shortcut in zero time, through the wormhole. We now give each mouth of the wormhole the same constant acceleration at a time $t = 0$ in the initial rest frame O (Fig. 5.5).

Viewed from the frame O the acceleration comes to an end at some time t and the wormhole mouths have each acquired a speed v, say. Viewed

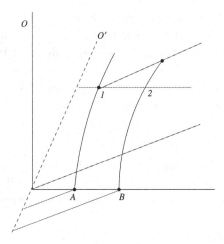

Fig. 5.5   The figure shows the world lines A and B of the two wormhole mouths in their initial rest frame O. Also shown dashed are the space and time axes of the frame O′ moving with speed v with respect to the frame O in the positive x-direction. The frame O′ is the final rest frame of the mouths A and B. Note that the events 1 and 2 at which the acceleration ceases are simultaneous in O but not in O′.

from the new rest frame of the wormhole mouths O′, mouth B comes to rest first at time $t'_b = \gamma(t - vx_b)$ followed by mouth A at the later time $t'_a = \gamma(t - vx_a)$ where we have used a Lorentz time transformation between frames. This is an example of the relativity of simultaneity. Therefore in O′ the wormhole connects different times with mouth A ahead of mouth B by an amount $T = v\gamma(x_b - x_a) = \gamma vs$ where s is the constant separation of the two mouths in O; their distance apart in frame O′ is γs. However, the time through the wormhole is unchanged: it is still zero. We therefore have the situation depicted in Fig. 5.6.

To complete our construction of a time machine we must arrange matters so that the time shift is greater than the light travel time between A and B. To do this we imagine bringing the mouths closer together slowly enough that the time shift is not significantly altered. Figure 5.7 shows the state of affairs if the time lapse as a result of this construction is larger than the speed of light journey the long way round. It is now possible to leave B on a timelike path, arrive at A, and return via the shortcut wormhole route to a time before setting off.

A similar construction can be carried out in principle with 'real' wormholes in curved spacetime. To move the wormholes physically now,

Fig. 5.6   As for Fig. 5.7 but viewed from the final rest frame O'. The axes of O are dashed here. The wormhole now joins an event on A to an earlier event on B. This is represented in the inset.

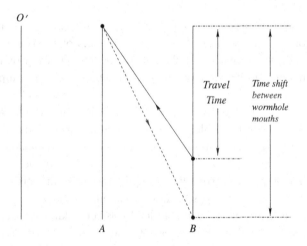

Fig. 5.7   The figure shows a closed-timelike path along which an observer can travel into their own past.

we could charge a wormhole mouth electrically, or shift it using gravity. Kinematically, observers stationed at the throats would experience the same changes to their spacetime under acceleration as did the Minkowski observers of our previous example, since the spacetime outside the throats is approximately flat. Notice that in the frame of an observer at one of the throats, the distance through the wormhole remains constant, so the travel time through the wormhole is unchanged. This corresponds to the picture of Fig. 5.3.

One can ask whether such a time machine is serviceable in the sense that it remains stable when normal matter moves into it. In fact, as long as the accelerations are sufficiently small, only a small change in the exotic matter content is required to keep the wormhole mouth open (Morris *et al.*, 1988).

An alternative way of achieving large timelike separations between the mouths of a wormhole is to take one mouth of the wormhole on a round trip journey at high speed (Thorne, 1994). This employs the well-known twin effect whereby one twin moves into the future of the other twin. Suppose the time separation achieved by this means is T. When the travelling wormhole mouth arrives back, it is then possible to step into it and come out a time T in the past (less the time it takes to travel through the wormhole, which can be relatively short).

5.8 Chronology protection

Time machines raise a number of problems that go beyond mere paradoxes. They all concern the well-known classic problems of interfering with the past, for example preventing one's own birth. The creation of wormholes leads to the possibility of time machines which seem to imply logical inconsistencies. Where in the argument have we gone wrong?

There are several possibilities: (i) Wormholes that would allow time travel to the past are unstable: as soon as such a wormhole is created light can travel through it an infinite number of times contributing a positive energy density that collapses the wormhole. (ii) Wormholes are possible but the laws of physics (somehow) forbid inconsistent actions (such as interfering with one's own past). This is called the chronology protection conjecture. (iii) Wormholes are forbidden by as yet unknown laws of physics. This is a popular get-out but is easier said than believed, because it is difficult to see where such laws might intervene in the discussion.

Chapter 6

Black Holes in Other Dimensions

In this chapter we shall try to give a flavour of some recent work on black holes. The subject has expanded enormously in recent years; the best we can do at the level of this book is to give some indication of the nature of these developments.

6.1 Why study other dimensions?

The fact that we live on an apparently four dimensional slice of spacetime does not imply that the universe has three dimensions of space and one of time (a '3 + 1' dimensional universe). There are two other possibilities. One is that the world has three large spatial dimensions, which form part of our everyday experience, plus a number of compact spatial dimensions which wrap round on a small scale and are only accessible to experiments that probe these smallest scales. The other possibility is that we live on a three dimensional hypersurface in a space with a greater number of dimensions. Some theories that attempt to embrace gravity and quantum theory require the existence of higher dimensions. One way to explore these theories is to look at what they have to say about black holes. For example, we can ask what they have to say about the problems we met in Section 3.20 on the origin of black hole entropy and the information paradox. Some properties of black holes, such as the laws of black hole thermodynamics, carry over to higher dimensions; others, such as the uniqueness theorems, do not. The properties of black holes in higher dimensions may therefore provide us with a way to probe the existence of these higher dimensions.

Earlier chapters of this book have been devoted to a small number of types of black holes. In fact, we noted in Section 3.6 that under certain assumption black holes can be characterised by just a few parameters: mass, angular momentum and charge. The results are known collectively as 'uniqueness theorems' and colloquially as 'no-hair theorems'. If we relax some of the assumptions in Section 3.6 then a new range of possibilities emerge.

If we allow a (negative) cosmological constant, which, as we shall see below, entails relaxing the condition that the spacetime must be asymptotically flat, it is no longer the case that the horizon must be topologically spherical. (A surface is topologically spherical if it can be continuously deformed into a sphere without tearing. Thus the horizon of a Kerr black hole is not geometrically a sphere – the equatorial and polar circumferences differ – but it is topologically spherical.) As an example, the metric

$$d\tau^2 = \left(\frac{r^2}{l^2} - \frac{2m}{r} + \frac{q^2}{r^2}\right) dt^2 - \left(\frac{r^2}{l^2} - \frac{2m}{r} + \frac{q^2}{r^2}\right)^{-1} dr^2 - r^2(d\theta^2 + d\phi^2)$$

(6.1)

is a solution of Einstein's equations with a cosmological constant $\Lambda = -3/l^2$ (see Mann, 2008). The event horizon at $r = r_+$, is given by $g_{00} = 0$ or

$$\frac{r_+^2}{l^2} - \frac{2m}{r_+} + \frac{q^2}{r_+^2} = 0.$$

For $r = r_+ = $ constant and $t = $ constant the angular part of the metric (6.1)

$$ds^2 = r_+^2(d\theta^2 + d\phi^2),$$

describes a torus with coordinates $0 \leq \theta < 2\pi$ and $0 \leq \phi < 2\pi$ (Fig. 6.1). Note that the torus is constructed from a region of the (flat) plane and is therefore intrinsically flat. It looks curved when embedded as an object in three dimensions because there we see the extrinsic curvature it inherits from the embedding space.

If we consider spacetimes with more than four dimensions, even restricting ourselves to asymptotic flatness, the number of possible stable

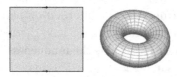

Fig. 6.1   Identifying the top and bottom edges and the left and right edges of the rectangle $0 \leq \theta < 2\pi, 0 \leq \phi < 2\pi$ gives the torus on the right. The intrinsic metric of the torus is flat, as can be seen on the left; the metric inherited from embedding the torus in 3d in the figure on the right is not flat: if the inner circumference is $2\pi(c - a)$ and the outer circumference is $2\pi(c + a)$ in the equatorial plane, then the inherited metric is $ds^2 = (c + a\cos\phi)^2 d\theta^2 + a^2 d\phi^2$.

black hole solutions becomes much larger. For five dimensions a lot is known about the various possibilities. As we go to even higher dimensions the range of possibilities increases and the fraction that have been explored decreases. The reason for this expansion in the numbers of black hole solutions is the increase in the number of degrees of freedom. In higher dimensions there are more geometrical structures of lower dimension that are candidates for horizon geometries: from black strings to black hypersurfaces (i.e. from one dimensional black objects to black objects with one less spatial dimension than the embedding space). In three space dimensions any rigid motion of an object about a fixed point can be described as rotation about an axis, where the axis can be specified as the normal to a plane. Thus, the addition of each pair of dimensions defines a new plane and hence a new possibility for an independent axis of rotation and a new type of rotating black hole.

So the problem with going to higher dimensions is that we get a plethora of solutions. It is not possible to provide a detailed review of the current situation. This chapter, unlike the previous ones, is intended to provide merely an overview of the way the research environment has changed by presenting a few, highly selective examples.

6.2 Anti-de Sitter black holes

If the cosmological constant Λ is positive, the vacuum solution to Einstein's field equations is the de Sitter metric. While this is the likely case cosmologically, from a theoretical point of view the anti-de Sitter spacetime, with negative $\Lambda = -3/l^2$, is more interesting. Anti-de Sitter is usually abbreviated as 'AdS'. We first look briefly at ways of writing the AdS metric and use these to present metrics for black holes in spacetimes that approach AdS at infinity.

One form of the the AdS metric is

$$d\tau^2 = \left(1 + \frac{r^2}{l^2}\right) dt^2 - \left(1 + \frac{r^2}{l^2}\right)^{-1} dr^2 - r^2 d\Omega^2, \qquad (6.2)$$

with, as usual, $d\Omega^2 = d\theta^2 + \sin^2 \theta d\phi^2$. The metric (6.2) is clearly not flat: it does not reduce to the Minkowski metric as $r \to \infty$. The rising value of the potential in the Newtonian approximation, $\Phi \sim g_{00} - 1 \propto r^2$, means that a free particle with finite energy and non-zero mass can never escape to infinity, but will oscillate about $r = 0$. As in Minkowski spacetime any such particle can be brought to rest by the equivalent of a Lorentz transformation, and any point can be chosen as $r = 0$, so there is no absolute rest and no absolute centre in the anti-de Sitter spacetime. In fact, the spacetime has constant negative curvature rather as a sphere has constant positive curvature, but with an imaginary radius. The corresponding real quantity l is often referred to as the radius of curvature of the AdS metric.

To obtain the Penrose diagram, so that we can see what infinity looks like in AdS spacetime, we define

$$x = \tan r/l.$$

The metric (6.2) becomes

$$d\tau^2 = l^2 \sec^2 x \left[d(t/l)^2 - dx^2 - \sin^2 x d\Omega^2 \right]. \tag{6.3}$$

Since the coordinate x ranges over a finite interval ($0 \le x \le \pi/2$ as r runs between 0 and ∞, the conformal metric (i.e. the metric we get by neglecting the overall factor on the right of (6.3)) is a cylinder in (t, x), with infinity corresponding to the surface of the cylinder at $x = \pi/2$.

We obtain a static black hole solution in a spacetime that is AdS at infinity by extension of the Schwarzschild solution:

$$d\tau^2 = \left(\frac{r^2}{l^2} + 1 - \frac{2m}{r} \right) dt^2 - \left(\frac{r^2}{l^2} + 1 - \frac{2m}{r} \right)^{-1} dr^2 + r^2 d\Omega^2, \tag{6.4}$$

with an event horizon at $r = r_h$ given by $g_{00} = 0$:

$$\frac{r_h^2}{l^2} + 1 - \frac{2m}{r_h} = 0.$$

As $r \to \infty$ the metric obviously tends to that of anti-de Sitter, equation (6.2) so far away from the horizon the geometry tends to AdS. As $l \to \infty$ the metric tends to Schwarzschild.

However, the situation is a lot more interesting: we can generate any number of horizon topologies for black holes in an asymptotically AdS spacetime. To see this with a minimum of algebra, we begin by looking at the AdS spacetime from a different point of view. We show that a spacetime of constant negative curvature (6.2) can be regarded as a four-dimensional surface embedded in a flat five-dimensional space. This is analogous to the way in which a two-dimensional sphere can be embedded in a three-dimensional Euclidean (flat) space.

We start from a five-dimensional flat space with metric

$$d\tau^2 = dT_1^2 + dT_2^2 - dX_1^2 - dX_2^2 - dX_3^2 \tag{6.5}$$

and consider the four-dimensional surface

$$T_1^2 + T_2^2 - X_1^2 - X_2^2 - X_3^2 = l^2. \tag{6.6}$$

The metric on this surface is the AdS metric (problem 6.1). To obtain the form (6.2) we parameterise (6.6) in terms of parameters t, r, θ, and ϕ

$$T_1 = l\sqrt{1 + (r/l)^2} \sin(t/l); \quad T_2 = l\sqrt{1 + (r/l)^2} \cos(t/l);$$

$$X_1 = r \sin\theta \cos\phi; \quad X_2 = r \sin\theta \sin\phi; \quad X_3 = r\cos\theta. \tag{6.7}$$

Problem 6.1. *Show that with the parameterisation (6.7) the surface defined by equation (6.6) inherits from 6.5 the AdS metric (6.2).*

To obtain our various black hole solutions we now define some new coordinates

$$\tilde{t} = \frac{T_2}{\rho}; \quad x_2 = \frac{X_2}{\rho}; \quad x_3 = \frac{X_3}{\rho},$$

with ρ as yet undefined. Substituting in (6.5) we get

$$d\tau^2 = \frac{\rho^2}{l^2}(d\tilde{t}^2 - dx_2^2 - dx_3^2) + \frac{l^2}{\rho^2}d\rho^2$$

$$+ \left[-dT_1^2 + dX_1^2 + 2\frac{d\rho}{\rho}(T_1 dT_1 - X_1 dX_1) + \frac{d\rho^2}{\rho^2}(-T_1^2 + X_1^2) \right].$$

We now choose ρ to ensure that the term in square brackets is zero. This is achieved by the choice

$$\rho = T_1 - X_1$$

which completes the definition of our new coordinates on the AdS hyperboloid. The AdS metric (6.2) can therefore be cast into the alternative form

$$d\tau^2 = \frac{\rho^2}{l^2}(d\tilde{t}^2 - dx_2^2 - dx_3^2) + \frac{l^2}{\rho^2}d\rho^2. \tag{6.8}$$

We can now get an AdS black hole (different from (6.4)) by modifying the metric (6.8) to

$$d\tau^2 = \frac{\rho^2}{l^2}\left[\left(1 - \frac{\rho_h^3}{\rho^3}\right)d\tilde{t}^2 - dx_2^2 - dx_3^2\right] + \frac{l^2}{\rho^2}\left(1 - \frac{\rho_h^3}{\rho^3}\right)^{-1}d\rho^2. \tag{6.9}$$

It is clear that this metric is AdS at infinity. (Compare (6.8) and (6.9) as $\rho \to \infty$.) It has a horizon at $\rho = \rho_h$, spanned by x_2 and x_3; furthermore the geometry of the horizon at $\rho = \rho_h = $ constant, $\tilde{t} = $ constant, is flat! Just as we obtained the torus in Fig. 6.1, by identifying the points x_2 and

$x_2 + 2\pi$ and the points x_3 and $x_3 + 2\pi$ (i.e. by regarding x_2 and x_3 as angles) we obtain a black hole with a toroidal horizon. The black hole resides in a spacetime that is locally AdS at large distances. By this we mean that the metric in any small region is (approximately) AdS. The spacetime is not globally AdS because two of the coordinates are still wound up into a torus at infinity. A torus is indistinguishable from flat space in any little patch, but it is distinguishable globally because one can travel back to any starting point without changing direction. In fact, by generalising this approach, one can generate asymptotically anti-de Sitter black holes with horizons that have any possible two dimensional topology (called spheres with handles or Riemannian surfaces).

6.3 Dimensions less than 4: The BTZ metric

Our main interest will be in spacetimes with more than three spatial dimensions, since these are not ruled out by observation. However, we shall start by looking at an example of a black hole in a spacetime with $2+1$ dimensions, that is just 2 spatial dimensions and one time dimension. The reason for this is that a lot of properties simplify. In particular, like Newtonian gravity, which just comprises an instantaneous potential, the $2+1$ theory has no propagating modes (i.e. no gravitational waves). If we restrict the spacetime to be asymptotically flat there are no black hole solutions. However, it turns out that there is a black hole metric satisfying Einstein's equations in three (spacetime) dimensions if we allow a negative cosmological constant, so the spacetime approaches anti-de Sitter at infinity. This is the Bañados, Teitelboim and Zanelli or BTZ black hole metric (Bañados *et al.*, 1993).

The BTZ metric for a black hole of mass $M = m/8G$ and angular momentum per unit mass a in $2+1$ gravity is

$$d\tau^2 = \frac{(r^2 - r_+^2)(r^2 - r_-^2)}{l^2 r^2} dt^2$$
$$- \frac{l^2 r^2}{(r^2 - r_+^2)(r^2 - r_-^2)} dr^2 - r^2 \left(d\phi - \frac{r_+ r_-}{l r^2} dt\right)^2 \quad (6.10)$$

where $-3l^{-2} = \Lambda$ is again the cosmological constant, and

$$r_\pm = l \left[\frac{m}{2}\left(1 \pm \sqrt{1 - \left(\frac{a}{l}\right)^2}\right)\right]^{1/2}. \quad (6.11)$$

As in the Kerr solution there is an outer event horizon at r_+ and an inner Cauchy horizon at r_- with an ergosphere from r_+ out to $r = lm^{1/2}$. The temperature of the hole can be calculated by an analogous argument to Hawking's original derivation sketched in Section 3.20 (and in various other ways). The result is

$$T = \frac{\hbar\kappa}{2\pi} = \frac{\hbar(r_+^2 - r_-^2)}{2\pi l^2 r_+}$$

and the entropy is

$$S = \frac{2\pi r_+}{4\hbar G}$$

which is proportional to the *perimeter* of the horizon. In $3 + 1$ dimensions we saw that the entropy of a black hole was proportional to the area of the horizon. Thus the entropy is not an extensive quantity: it is not proportional to the volume of the hole. Similarly we find for the BTZ black hole that the entropy is not an extensive quantity: it is proportional to the perimeter of the hole, not to the area of the 2d space occupied by the hole.

Since the BTZ black hole has no propagating degrees of freedom (i.e. no excitable modes which could contribute to the density of states) one might wonder where the entropy of the black hole originates. The answer seems to be that there are dynamical degrees of freedom confined to the boundary of anti-de Sitter space at infinity. It is not clear how to quantise these modes to enable the density of states to be computed directly, but it turns out that there are ways of counting states in the resulting type of field theory that does not depend on the quantum details, and the result agrees with the Hawking formula (see Section 6.7.4).

Problem 6.2. *In the absence of angular momentum ($a = 0$), the BTZ black hole is the $2 + 1$ dimensional analogue of the Schwarzschild solution for a spacetime that approaches the anti-de Sitter metric at spatial infinity (6.4). For $a = 0$, show that $r_+ = l\sqrt{m}$, $r_- = 0$ and that the BTZ metric can be written as*

$$d\tau^2 = \left(-m + \frac{r^2}{l^2}\right) dt^2 - \frac{dr^2}{-m + \frac{r^2}{l^2}} + r^2 d\phi^2.$$

Strictly speaking the result of problem 6.2 does not give a relation between the horizon radius and the mass because we have not shown that m is the black hole mass; however, this is as far as we can go with elementary

methods because the definition of mass is quite subtle in spacetimes that
are not asymptotically flat.

Problem 6.3. *The Poisson equation for the Newtonian gravitational
potential φ of a point mass at the origin in two space dimensions with
a cosmological constant is*

$$\frac{1}{r}\frac{d}{dr}\left(r\frac{d\varphi}{dr}\right) = \frac{GM}{r}\delta(r) + \left(\frac{2}{l}\right)^2$$

*where $\delta(r)$ is the Dirac delta function. Show that with a judicious choice of
integration constants the Newtonian potential is related to the BTZ metric
by $g_{00} = \varphi$.*

6.4 Higher dimensional black holes in general relativity

Turning now to higher dimensions the first thing we note is that the
restrictions on the classical theory are not as constraining as in lower
dimensions. The first aspect is that there is more flexibility in the field
equations (the relation between curvature and matter content). Other
aspects are the uniqueness ('no hair') theorems and the topology ('shape')
of a black hole. In addition, a lot of interest has focussed on black holes
in spacetimes with a cosmological constant, particularly in AdS, which is
a vacuum solution of Einstein's equations with a negative cosmological
constant. The principle reason for this interest is that there appears to be
a correspondence, first noted by Malcedena (1998), between a certain class
of field theories (conformal field theories or CFTs) on the boundary of AdS
and gravity theories in the interior (called the 'bulk' in the literature). Now,
the quantum version of gravity cannot be a normal field theory, because
such a theory cannot account for the non-extensive nature of black hole
entropy. But this AdS/CFT correspondence, or its generalisation to other
spacetimes in what is known as the holographic principle, if true, tells us
that gravity can be quantised as a *field theory on the boundary spacetime*
and the results translated back, through this correspondence, to the interior.

6.4.1 *Schwarzschild–Tangherlini black holes*

The simplest generalisation of a Schwarzschild black hole, in asymptotically
flat spacetime, to d spacetime dimensions is obtained by joining the $r - t$
plane to a higher dimensional sphere. These solutions are usually called

simply Schwarzschild black holes. The metric

$$d\tau^2 = fdt^2 - f^{-1}dr^2 - r^2 d\Omega_{d-2}^2 \qquad (6.12)$$

where

$$d\Omega_{d-2}^2 = d\theta_1^2 + \sin^2\theta_1 d\theta_2^2 + \sin^2\theta_1 \sin^2\theta_2 d\theta_3^2 + \cdots.$$

was shown by Tangherlini to satisfy Einstein's vacuum equations with

$$f(r) = 1 - \frac{16\pi G_d M}{(d-2)\Omega_{d-2} r^{d-3}}$$

where G_d is Newton's constant in d dimensions. The quantity $d\Omega_{d-2}^2$ is the metric of a unit sphere in $d-2$ dimensions with area

$$\Omega_{d-2} = \int_0^{2\pi} \int_0^{\pi} \cdots \int_0^{\pi} \sin^{d-3}\theta_1 \sin^{d-4}\theta_2 \cdots \sin\theta_{d-3} d\theta_1 \cdots d\theta_{d-2}$$

$$= \frac{2\pi^{(d-1)/2}}{\Gamma\left(\frac{d}{2} - \frac{1}{2}\right)}.$$

The gamma function is defined by the recurrence relation

$$\Gamma(z+1) = z\Gamma(z)$$

with $\Gamma(1) = 1$ and $\Gamma(1/2) = \sqrt{\pi}$.

To show how the connection with Newtonian gravity can be made (and hence to justify the identification of the parameter M as the mass) we note that with the Newtonian approximation for the metric, $g_{00} = 1 - 2\Phi$ (Section 2.3.5), the Einstein field equations reduce to

$$\nabla_{d-1}^2 \Phi = -16\pi G_d \left(\frac{d-3}{d-2}\right) T_{00},$$

where ∇_{d-1}^2 is the Laplace operator in $d-1$ spatial dimensions. The mass (called the ADM mass in the literature) is then given by

$$M = \int T_{00} dx^{d-1} = -\left(\frac{d-2}{d-3}\right) \frac{1}{16\pi G_d} \int \frac{\partial\Phi}{\partial n} dx^{d-2},$$

where the first integral is over all space and the second is over a spacelike surface at infinity with normal $\mathbf{n}$. The ADM mass is the relativistic generalisation of the usual expression $M = \int \rho dV$.

The metric (6.12) is in fact the only static vacuum solution in asymptotically flat spacetimes, just as in four dimensions. A similar extension of the Reissner–Nordström metric to higher dimensions can be made for charged black holes. Apart from the higher dimensionality of the event horizon (a $d - 2$ dimensional sphere), and, of course, of the surrounding space, the properties of the metrics are similar to the four dimensional case. In d spacetime dimensions the charged black hole metric with mass M and charge Q is

$$d\tau^2 = f dt^2 - f^{-1}dr^2 - r^2 d\Omega_{d-2}^2 \qquad (6.13)$$

where

$$f(r) = 1 - \frac{16\pi G_d}{(d-2)\Omega_{d-2}} \frac{M}{r^{d-3}} + \frac{8\pi G_d}{(d-2)(d-3)\Omega_{d-2}} \frac{Q^2}{r^{2(d-3)}} \qquad (6.14)$$

$$= \left[1 - \left(\frac{r_1}{r}\right)^{(d-3)}\right] \left[1 - \left(\frac{r_2}{r}\right)^{(d-3)}\right] \qquad (6.15)$$

where r_1 and r_2 are solutions of $f(r) = 0$. The larger of these two solutions is the outer event horizon. In the extremal case $r_1 = r_2$.

6.4.2 *Spinning black holes: The Myers–Perry metrics*

When we turn to analogues of spinning black holes things become a lot more complicated. The analogues of the Kerr metric are the Myers–Perry black holes (Meyers & Perry, 1986). If we restrict the rotation to a single plane the metric in $d \geq 4$ spacetime dimensions is

$$d\tau^2 = dt^2 - \frac{\mu}{r^{d-5}\rho^2}(dt - a\sin^2\theta d\phi)^2 - \frac{\rho^2}{\Delta}dr^2 - \rho^2 d\theta^2$$
$$- (r^2 + a^2)\sin^2\theta d\phi^2 - r^2\cos^2\theta d\Omega_{d-4}^2$$

where

$$\rho^2 = r^2 + a^2\cos^2\theta$$

and

$$\Delta = r^2 + a^2 - \frac{\mu}{r^{d-5}}. \qquad (6.16)$$

The mass and angular momentum are given by

$$M = \frac{(d-2)\Omega_{d-2}}{16\pi G_d}\mu$$

and

$$J = \frac{2}{d-2}Ma,$$

where μ is a constant.

There are a number of similarities between the properties of the solution in higher dimensions but also some differences. The differences arise because the gravitational term $\mu/r^{(d-5)}$ in the expression for Δ (6.16) depends on dimension, whereas the centrifugal term, a^2, does not. Since the event horizon lies at (the largest real root of) $\Delta = 0$ the properties of the horizon depend on the dimension.

Recall that for $d = 4$ the solution has an event horizon up to the extremal value of rotation at $a \leq m$, beyond which it is a naked singularity. For $d = 5$, $\Delta = 0$ has a real solution, and hence a horizon, up to $\mu = a^2$. However, the area of the limiting case vanishes and it is a naked singularity. For $d \geq 6$, Δ is negative at small r and positive at large r. It therefore always has a positive real root independent of the value of a. This means that, unlike the lower dimensional cases, the metric has a horizon for arbitrarily large values of angular momentum. As one might expect, the horizon flattens as the spin increases. In the limit the hole becomes a 'black membrane' with an infinite flat horizon geometry. However, it would appear that these 'ultraspinning' black holes are unstable to small perturbations for $d \geq 6$.

Problem 6.4. *Show that the Myers–Perry metric reduces to the Kerr metric for $d = 4$ with $GM = m$ and $J = Ma$.*

6.4.3 *Topological black holes*

In Sections 6.2 and 6.1 we gave examples of non-spherical black holes that are asymptotically AdS. We saw that in four-dimension we can characterise the horizon geometry of an asymptotically AdS black hole completely. In the previous section we gave an example of a non-spherical event horizon which can arise in higher dimensions. In this section we shall outline briefly some of the other event horizon topologies that can arise in spacetimes in higher dimensions that are asymptotically flat. A simple example that can be obtained by a different extension of the Schwarzschild solution from that in Section 6.4.1 is:

$$d\tau^2 = f dt^2 - f^{-1} dr^2 - r^2 d\Omega^2 - dz^2 \qquad (6.17)$$

in which the fifth dimension is flat. This object has a cylindrical event horizon (because it is infinitely extended in the z-direction, with r_h independent of z). It is called a black string. The metric satisfies Einstein's equations with $f(r) = 1 - r_h/r$, as in the Schwarzschild case in four dimensions. From the four dimensional perspective, for example for orbits in the the hyperplane $z = $ constant, the mass of the source is $GM = r_h/2$. This is not the mass of the string itself, which is infinite, and consequently the spacetime is not asymptotically flat.

We now imagine taking the string, or a finite piece of it, and bending it into a loop. We guess that the loop can be prevented from collapsing under its own gravity by spinning it up. This suggests that in five dimensions (and higher) there will exist black rings with horizons that take the form (sphere) × (circle). So, for example, in five dimensions the horizon would be the product of an ordinary 2-sphere with a circle. This is quite difficult to picture, so imagine first a great circle on the 2-sphere (a section of the sphere at constant azimuthal angle, ϕ). Now take the product with the inner edge of the ring, with coordinate ψ, say. This gives us a torus, the product of a circle (a 1-sphere) with a circle, spanned by the angles (ϕ, ψ), which we *can* picture. It turns out that there is a range of values of mass and angular momentum for which there exits two solutions of this type: a fat black ring and a thin black ring. The thin rings are related to the black string solutions in the limit of large radii; the fat rings are more like the Myers–Perry black holes with a hole punched through. We can conclude that in dimensions ≥ 5, there is more than one black hole with a given value of mass and angular momentum. Since these are the only conserved quantities characterising the hole, the uniqueness results do not carry over from four dimensions.

We can however restrict the topology somewhat if we impose the dominant energy condition (Section 5.5.1). Under this condition on the matter content, an event horizon in an asymptotically flat spacetime must (roughly speaking) have positive scalar curvature. This rules out tori and restricts the topology to products of spheres.

6.5 Black holes in string theories

String theory is a fundamental theory of matter and fields in which the basic entities are envisaged to be not zero-dimensional point-like particles, but one-dimensional objects like strings. Particles emerge as excitations of the string, that is, as waves on the string. The lowest energy excitations (of relativistic strings) correspond to massless particles; the higher energy excitations correspond to particles with very large masses which are not observed at normally accessible energies. At low energies therefore string theory looks like an ordinary quantum field theory. This theory is called supergravity. We make three observations. First, there is more than one string theory so there is more than one type of supergravity. Second, the super in supergravity derives from the inclusion in the theory of supersymmetric partners to the particles with which we are normally familiar. In a supersymmetric theory, to every boson there is a corresponding fermion, so for example, the spin-1 photon is partnered by the spin-$\frac{1}{2}$ photino. The supersymmetric partners are assumed to be sufficiently massive that they are not yet observed in collider experiments. Finally, the gravity in supergravity is not one of the the fundamental fields, but also appears as a low energy excitation of the string. In this sense spacetime is an emergent property of string theory.

The great attractions of supersymmetric string theory are: supersymmetry restricts the form of the interactions that can occur and there are few free parameters with the promise that the four string theories are special cases of a unique (to be found) overarching theory of everything, called M-theory.

However, supersymmetric string theories and M-theory require for consistency that spacetime has 10 or 11 dimensions. The interest in higher dimensional spacetimes arises in the main therefore from string theory and its low energy limit, supergravity.

But the appeal of higher dimensions dates back to well before the advent of string theories with attempts to construct a geometrical theory of electromagnetism in five dimensions in which the extra components of the metric arising from the fifth dimension represent the electromagnetic field. This is the Kaluza–Klein theory (Section 6.5.1). The fact that (in this theory) we cannot move around in the extra dimension, but can experience it only in the form of electromagnetic interactions, can be understood if that dimension is wrapped into a sufficiently small circle. If the circle has a length λ then it would be accessible only at energies exceeding hc/λ.

The original interest in black holes in string theories came from the attempts to understand the origin of black hole entropy. Since black hole thermodynamics involves quantum mechanics, it would appear that we need a quantum theory of gravity to address the issue of entropy. That being so, we should not be surprised that we cannot use the effective classical low energy theory of gravity to count degrees of freedom. At first it was thought that quantum gravity would be irrelevant for classical black holes, because the black hole horizon is a much larger length scale than the Planck length. However, this does not appear to be the case. In a sense, in a black hole the quantum strings spread out over a horizon scale and quantum properties of the string excitations, including gravity, are important (Section 6.7.5).

String theory provides a second way of hiding the extra dimensions. The theory contains not only strings but also the higher dimensional objects to which the end points of open strings are attached. (See Section 6.5.2.) These are called 'branes'. So we have 0-branes or points, 1-branes, or strings and the generalisations to p-branes. Particularly confusing, at first, is the fact that the relevant sorts of branes are called D-branes, which has nothing to do with their dimension, but with the way in which the strings attach: so, if we want to be specific, we can have Dp-branes. (D stands for Dirichlet, which is a term generally associated with fixed endpoints and unconstrained slopes.) We have already encountered black strings; we can have black branes as well (Section 6.5.2). In this approach our universe appears as a four-dimensional brane in a higher dimensional spacetime. Because we live in an apparently three-dimensional world the extra dimensions must be hidden from direct view; but they manifest themselves as fields in our four-dimensional brane. We show this next.

6.5.1 *Kaluza–Klein black holes*

To understand the Kaluza–Klein approach we construct a five dimensional metric G_{mn}, $(m, n = 0, \ldots, 4)$ from the four dimensional metric $g_{\mu\nu}$, $(\mu, \nu = 0, \ldots, 3)$ with the help of a scalar field ϕ and a vector field A_μ. In block matrix form we put

$$G_{mn} = \left(\begin{array}{c|c} g_{\mu\nu} - \phi A_\mu A_\nu & -\phi A_\mu \\ \hline -\phi A_\nu & -\phi \end{array} \right). \tag{6.18}$$

From a five-dimensional point of view G_{mn} is a pure gravity vacuum solution of Einstein's equations. From the four dimensional point of view $g_{\mu\nu}$ is a solution of Einstein's equations with an electromagnetic field $F_{\mu\nu}$ (derived from the vector potential A_μ with $F_{\mu\nu} = \partial_\mu A_\nu - \partial_\nu A_\mu$) that satisfies Maxwell's equations. (In terms of the electric field **E** and the magnetic field **B** we have $E_i = F_{0i}$ and $B_1 = F_{23}$ etc; the Einstein equations in the fifth dimension turn out to be equivalent to Maxwell's vacuum equations in four dimensions for **E** and **B**.) The theory (in this form) also contains an extra scalar field ϕ which is not identified with a known physical field. (But, of course, this is not a complete theory since it does not contain any of the other known particles beyond the photon; it is illustrative of the approach.)

Consider again the five dimensional Schwarzschild metric (6.17) which we rewrite as

$$d\tau_5^2 = G_{mn}dx^m dx^n = (1 - 2m/r)dt'^2$$
$$-(1 - 2m/r)^{-1}dr^2 - r^2 d\Omega^2 - dz'^2. \tag{6.19}$$

We give the string a momentum in the fifth dimension by a Lorentz transformation. We shall see that from the four-dimensional point of view this is equivalent to introducing an electric charge on the string. The idea that momenta in the higher dimensional compact space is seen as a charge in lower dimensions will be important in evaluating the black hole entropy.

We now make a Lorentz transformation of the metric (6.19) to a frame moving with speed v along the $z-$ axis:

$$t' = \gamma(t - vz)$$
$$z' = \gamma(z - vt)$$

where $\gamma = (1 - v^2)^{-1/2}$. We obtain

$$d\tau_5^2 = -\left(1 + \frac{2mv^2\gamma^2}{r}\right)\left(dz - \frac{2m\gamma^2 v}{r + 2mv^2\gamma^2}dt\right)^2$$
$$+ \left(1 + \frac{2mv^2\gamma^2}{r}\right)^{-1/2}(fdt^2 - f^{-1}dr^2 - R^2 d\Omega^2) \tag{6.20}$$

where

$$R^2 = r^2\left(1 + \frac{2mv^2\gamma^2}{r}\right)^{1/2}, \quad f = \frac{1 - 2m/r}{(1 + 2mv^2\gamma^2/r)^{1/2}} \tag{6.21}$$

From the dz^2 term, by comparison with (6.18), we identify

$$\phi = 1 + \frac{2mv^2\gamma^2}{r}.$$

Since none of the metric coefficients in (6.20) depend on z we can identify the four-dimensional form

$$d\tau^2 = fdt^2 - f^{-1}dr^2 - R^2 d\Omega^2 \tag{6.22}$$

up to an overall factor of $\phi^{-1/2}$. From the point of view of Einstein's equations the five dimensional metric is a vacuum solution. But by comparison of (6.20) with the G_{05} component in (6.18) we can identify the Coulomb potential

$$A_0 = -\frac{2m\gamma^2 v}{r + 2mv^2\gamma^2}.$$

The four dimensional metric arises from an energy-momentum tensor corresponding to an electric field derived from this potential and the additional scalar field ϕ.

This can also be seen from the fact that we can write the action in the two equivalent forms

$$\frac{1}{\lambda} \int d^4x dz \sqrt{-\det G_{mn}}\, R^{(5)}$$

$$= \int d^4x \sqrt{-g}(R - 2(\nabla\phi)^2 - e^{-2\sqrt{3}\phi}F^2) \qquad (6.23)$$

where λ is the circumference of the fifth dimension. On the left of (6.23) we have the Einstein action (Section 1.16) for vacuum gravity in five dimensions; on the right we have the action written in four dimensions but now with additional terms representing a scalar matter field ϕ and an electromagnetic field F^2. If the fifth dimension is wound up sufficiently tightly that we do not have access to it, the additional dimension will be detectable only through the electric (and scalar) fields.

The 4-metric represents a black hole with an event horizon ($f = 0$) at $r = 2m$ and a singularity at $r = 0$. To identify the mass of the hole we look at the asymptotic behaviour of f and compare it with the Schwarzschild solution. Since $r \sim R$ as $r \to \infty$ (see equation (6.21)),

$$f \sim 1 - \frac{2m + mv^2\gamma^2}{r} \sim 1 - \frac{2GM}{R}$$

where G is the four-dimensional Newton constant. Thus

$$GM = \frac{m}{2}(\gamma^2 + 1).$$

To get the charge we look at the behaviour of the Coulomb potential as $r \to \infty$:

$$A_0 = \frac{-2m\gamma^2 v}{r + 2mv^2\gamma^2} \sim -\frac{Q}{R} \qquad (6.24)$$

which leads to

$$Q = 2mv\gamma^2.$$

This is a key result: what appears as electric charge in four dimensions arises from momentum (mv) in five dimensions.

6.5.2 *Black branes*

The simplest solution of Maxwell's equations in electromagnetism is that of a point charge. In string theories, and hence in supergravity theories, as well as point charges (zero dimensional objects) it turns out that higher dimensional charged entities must also exist. These are called branes, short for membranes. The simplest spacetimes in supergravity theories will contain static charged branes, analogues to the Coulomb field of a point charge at rest and the gravitational effect of its mass and electric field.

Now, if we try to make a point charge with non-zero mass in classical relativity, we actually end up with a charged black hole. Equally, if we imagine putting a number of point masses together to create a mass distribution of sufficient density, we again end up with a black hole. Analogously, in supergravity we can create (only theoretically so far) black branes, in addition to the black strings we have already come across, and we can create black holes from multiple coincident branes.

For some black holes created in this way we can identify and count the degrees of freedom, hence we can calculate the entropy (Section 6.7.4).

6.5.3 *Warped branes*

Generally, when we consider wrapping up the extra dimensions by identifying points, we take the metric coefficients to be independent of the coordinates in the extra dimensions. If we want to take seriously the braneworld picture where the extra dimensions are not wrapped up, we have to consider the way gravity behaves away from the brane. This means we have to allow the metric to be a function of the coordinates orthogonal to the brane. A popular model is the five-dimensional Randall–Sundrum (RS) metric

$$d\tau^2 = e^{-|z|/l}(dt^2 - dx_1^2 - dx_2^2 - dx_3^2) - dz^2.$$

This metric corresponds to a five-dimensional AdS spacetime with radius of curvature l with a discontinuity in the curvature at $z = 0$ representing the surface of the brane. The hypersurfaces $z = $ constant are conformally flat; the conformal factor, here $e^{-|z|/\lambda}$, is known as the warp factor. The metric on the brane at $z = 0$ is flat spacetime, so it has the usual Minkowski vacuum. Although the induced metric of the four-dimensional world is flat, from the four-dimensional perspective the energy momentum tensor is non-zero! (It is proportional, in fact, to the Minkowski metric. It arises from the discontinuity in the five-dimensional curvature on the brane arising from the discontinuity in the z-derivative of the conformal factor.)

What has this to do with black holes? We have just seen that the five-dimensional vacuum geometry can appear as matter from the brane perspective. It turns out that a black hole in the bulk may look like radiation in the brane. Furthermore it is conjectured that a classical black hole solution in five dimension represents a solution of quantum gravity in the brane, by analogy with the AdS/CFT correspondence. Thus we might get the following rather appealing picture of black hole evaporation. Let the brane intersect the horizon of a five-dimensional black hole that is moving away from it. As the hole recedes, the intersection of the horizon shrinks, corresponding to the evaporation of the hole. From the persective of the brane we would obtain a fully quantum corrected picture of black hole evaporation consistently incorporating the effect of the radiation itself on the black hole (Gregory, 2009).

6.6 Stability

The black string solution we found in Section 6.4.3 turns out to be unstable. Problem 6.5 indicates why this is to be expected. This suggests also that many of the higher dimensional black objects will be unstable. The interesting feature of the black string instability is the way it evolves. A rather heroic numerical calculation leads to the conclusion that the string horizon develops large scale ripples, each of which ripples further and so on in a fractal pattern. The final state is a massless naked singularity providing a generic counter-example to the cosmic censorship hypothesis.

Problem 6.5. (a) *Show that the entropy of a five-dimensional black hole is given by*

$$S_{bh} = \frac{\pi^2 r_5^3}{2G_5}$$

where G_5 is the five-dimensional Newton constant and r_5 is the horizon radius. (Hint: the horizon is a 3-sphere; see Section 6.4.1)

(b) *Show that the radius of the horizon is given in terms of the mass of the hole by*

$$r_5^2 = \frac{8G_5 M}{3\pi}.$$

(c) *Show that the entropy of the black string (Section 6.4.3) (as a gravitating object in four dimensions) is given by*

$$S_{bs} = \frac{\pi r_+^2}{G} = 4\pi G M^2.$$

Given that $G = G_5/L$ for a string of length L wrapped round the compact fifth dimension, show that for sufficiently large L the black hole will be entropically favoured over the black string of the same mass, and hence that the black string solution is expected to be unstable.

6.7 Black holes in alternative theories of gravity

6.7.1 *Pure gravity alternatives: Lovelock theory*

The requirement that the field equations of general relativity reduce to the Newtonian equations that give exactly the inverse square law of gravitational force uniquely determines the form of the field equations in four spacetime dimensions (1.3.11). In higher dimensions additional terms non-linear in the curvature can be added to the action. In Einstein–Gauss–Bonnet gravity an additional term, non-linear in the curvature, is added to the Einstein–Hilbert Lagrangian (equation 1.16), so the field equations are derived from variation of

$$\int d^d x \sqrt{-g}(R + 2\Lambda + R^2 - 4R_{\mu\nu}R^{\mu\nu} + R_{\mu\nu\rho\sigma}R^{\mu\nu\rho\sigma}).$$

In four dimensions the additional non-linear terms do not contribute to the field equations because they reduce to a pure divergence (and can therefore be integrated to a vanishing surface term by the divergence theorem). In higher dimensions this is no longer true.

In dimension d Lovelock showed that we can add $k = (d-1)/2$ (d odd) or $k = d/2 - 1$ (d even) terms involving higher powers of the curvature, up to the k th power, with arbitrary coefficients. This is called Lovelock gravity. By placing restrictions on the coefficients of the extra terms in the Lagrangian one can obtain black hole solutions of Lovelock gravity. Apart from the fact that higher order curvature terms arise naturally in various approaches to quantum gravity, there has been much interest in Lovelock theory in the context of brane worlds, in particular in the application to branes of more than one dimension less than that of the bulk spacetime.

6.7.2 *Alternative matter*

As an alternative to changing the gravitational part of the Lagrangian as in the previous section, we can consider various modifications to the matter content. This approach has a long history in relation to attempts to rid classical electrodynamics of the singularities associated with point charges. The best known of these is the Born–Infeld theory which incorporates an upper limit for the electric field by analogy with the limiting speed of light in relativity. The Lagrangian density is

$$L = 1 - \sqrt{1 - (E^2 - B^2)/b^2 - (E.B)^2/b^4},$$

where b is a constant. This may be written elegantly in the form of a determinant

$$L = 1 - \sqrt{-det(g_{\mu\nu} + b^{-1}F_{\mu\nu})}$$

for a general metric $g_{\mu\nu}$. The square root structure is analogous to the way in which matter fields appear on branes in dimensionally reduced supergravity theories. This has led to the investigation of black holes in Einstein–Born–Infeld theory, whereby the Born–Infeld electromagnetic field is coupled to gravity through Einstein's equations. This modifies the properties of charged black holes, usually by removing or ameliorating singularities.

Problem 6.6. *A charged black hole solution to the Einstein–Born–Infeld equations is (Rasheed, 2008)*

$$d\tau^2 = \left(1 - \frac{2m(r)}{r}\right) dt^2 - \left(1 - \frac{2m(r)}{r}\right)^{-1} dr^2 - r^2(d\theta^2 + \sin^2\theta d\phi^2),$$

where the mass is given by

$$M = m(r) + \frac{1}{b^2} \int_r^\infty dx\left(\sqrt{x^4 + b^2Q^2} - x^2\right).$$

Let r_+ denote the radius of the (outer) event horizon (assuming it exists). The expression for the surface gravity of the horizon (problem 4.1) *reduces to*

$$\kappa = -\frac{1}{2}\frac{dg_{00}}{dr}\bigg|_{r=r_+}.$$

Show that the surface gravity is explicitly

$$\kappa = \frac{1}{2r_+} - \frac{1}{b^2 r_+}\left(\sqrt{r_+^4 + b^2 Q^2} - r_+^2\right).$$

Given that the electric potential on the horizon, area A, is

$$\Phi_h = \int_{r_+}^{\infty} dr \frac{Q}{\sqrt{r^4 + b^2 Q^2}}$$

verify that the black hole satisfies the first law of black hole mechanics:

$$dM = \frac{\kappa}{8\pi}dA + \Phi_h dQ.$$

An alternative approach is to ignore the matter content and focus on metrics with interesting properties; after all, we have just seen that the relation between matter content and metric is by no means fixed since we do not have a definitive set of field equations. The Lifshitz theory (often called Horava–Lifshitz theory) relinquishes Lorentz invariance for the sake of renormalisability of the quantum theory (meaning that quantum corrections to classical behaviour do not introduce an infinite number of arbitrary parameters). The argument here is that Lorentz invariance is only verified experimentally below some large but finite energy, above which the symmetry may be broken. One can then retain second order time derivatives in the field equations (essential for a sensible quantum theory that preserves probability), but have different orders of space derivatives (so time and space appear on different footings).

An example of a Lifshitz metric in $3 + 1$ dimensions is

$$d\tau^2 = \frac{r^{2z}}{l^{2z}}dt^2 - \frac{l^2}{r^2}(dr^2 + dx_1^2 + dx_2^2)$$

where z is a constant and l^{-2} is the cosmological constant (to within a numerical factor). Scaling time by $t \to \alpha^z t$, for some constant α, and space by $(r, x_1, x_2) \to (\alpha^{-1}r, \alpha^{-1}x_1, \alpha^{-1}x_2)$ leaves the metric unchanged.

Because of the nice quantum properties, there is some interest in finding black hole solutions that asymptote Lifshitz metrics as $r \to \infty$.

Problem 6.7. *An asymptotically Lifshitz black hole metric in* $2 + 1$ *dimensions is given by*

$$d\tau^2 = \frac{r^{2z}}{l^{2z}}\left(1 - \frac{ml^2}{r^2}\right)dt^2 - \frac{l^2}{r^2}\left(1 - \frac{ml^2}{r^2}\right)^{-1}dr^2 - r^2 d\phi^2.$$

Show that $z = 1$ *produces the BTZ black hole. In general (for a diagonal metric with* $g_{00} \neq g_{11}^{-1}$*) the expression for the surface gravity (compare problem 4.1) becomes*

$$\kappa = \frac{1}{2}\left[\sqrt{-g^{00}g^{11}}\left|\frac{dg_{00}}{dr}\right|\right]_{r=r_+}.$$

For general z *find the Hawking temperature from the surface gravity. The mass of the black hole in the case* $z = 3$ *is given by (Myung, 2012)*

$$M = \frac{r_+^4}{4G_3 l^4},$$

and the entropy is $S = A/G_3$ *where* A *is the area (i.e. perimeter) of the hole. Show that*

$$dM = T_{bh}dS_{bh}.$$

6.7.3 *AdS/CFT correspondence*

The entropy of a physical system is related to the number of microstates, or number of degrees of freedom, Ω, by

$$S = k_B \log \Omega.$$

In quantum theory the entropy is an extensive quantity, i.e. the entropy is proportional to the volume: if there are Ω states in a volume V there are Ω^2 in a volume $2V$. Once we admit gravity into the picture this simple argument breaks down.

Consider a spherical mass just outside its Schwarzschild horizon, on the verge of becoming a black hole. If its gravity were described by a conventional quantised field, the entropy of the system would be proportional to the volume of the sphere. Now add a tiny spherical shell of matter, tipping the system into a black hole. Its entropy, which should have changed by only a small amount, is now proportional to the area of the hole. A quantum field description of gravity would provide too many degrees of freedom to be compatible with this conclusion.

This leads to the idea that in general the quantum behaviour of gravitating systems must somehow be expressed holographically. It is impossible to discuss the holographic principle at the level of this book, so one illustration will have to suffice. This will be an example of what is known as the AdS/CFT correspondence. 'AdS' we have already met (Section 6.2). 'CFT' stands for conformal field theory, which means a theory that is unchanged in its physical predictions when the metric is multiplied by an overall function of the coordinates. Maxwellian electrodynamics in a vacuum (no charges or currents) is a (classical) conformal theory.

Our illustration involves a three-dimensional brane (a D3-brane) in 10 dimensions, which we shall describe (in outline) in terms of a field theory on the boundary of the spacetime. We shall indicate how the degrees of freedom of a bulk configuration are numerically equal to those of the boundary theory. Our starting point is the metric (Becker *et al.*, 2007)

$$d\tau^2 = \frac{r^2}{R^2}(f(r)dt^2 - dw^2 - dx^2 - dy^2) - \frac{R^2}{r^2}f^{-1}(r)dr^2 - R^2 d\Omega_5^2 \quad (6.25)$$

where

$$f(r) = 1 - \left(\frac{r_0}{r}\right)^4$$

with r_0 a constant. The metric tends to that of $AdS_5 \times S^5$ as $r \to \infty$. This means that five of the ten dimensions are AdS and the other five form a five-dimensional sphere. Each of the two subspaces here has a radius R. The metric describes a D3 brane with a charge N (see Section 6.7.4), or a set of N D3 branes, where

$$N^2 \sim \frac{R^8}{G_{10}}. \quad (6.26)$$

One can think of the corresponding field theory as describing the deformations of the D3 brane in the AdS spacetime, since such deformations are described by functions on the brane giving the orthogonal displacement of the brane at each point into the surrounding bulk space. The field theory contains N^2 scalar fields.

The temperature of the horizon is given by requiring periodicity in the Euclidean metric (Section 4.5.5). The horizon is at $r = r_0$. To obtain the near-horizon metric we put

$$r = r_0 + \frac{r_0}{R^2}\rho^2; \quad \theta = 2it\frac{r_0}{R^2}.$$

We obtain

$$d\tau^2 = \rho^2 d\theta^2 + d\rho^2 + \cdots .$$

Thus, making θ periodic with a period of 2π is equivalent to making *it* periodic with a period $\pi R^2/r_0$, and therefore the temperature of the horizon is given by $k_B T = r_0/(\pi R^2)$.

The entropy is given by

$$S = \frac{A}{4G_{10}} = \frac{(r_0/R)^3 \times R^5 \Omega_5}{4G_{10}}$$

where G_{10} is the Newtonian gravitational constant in ten dimensions and A is the area of the horizon. Using (6.26) we obtain the entropy

$$S \sim N^2 T^3.$$

For the corresponding field theory on the boundary we get a contribution to the entropy proportional to T^3 for each of the N^2 fields (just like blackbody radiation for photons) and hence an entropy $S \sim N^2 T^3$. This illustrates, albeit to a rather limited extent, a connection between quantum fields on the boundary and quantum gravity in the 'bulk'. In fact, in the way we have presented this there is a certain amount of sleight of hand. We have counted the states for free (non-interacting) quantum fields, but if there is no interaction (no gravity) we should not have a black hole at all. A more careful treatment of the approximations is required to make the derivation rigorous.

Problem 6.8. *Show that the metric in equation (6.25) is asymptotic to $AdS_5 \times S^5$ as $r \to \infty$.*

Problem 6.9. *Using the method of this section show that the temperature of a Reissner–Nordström black hole in d-spacetime dimensions is*

$$T = \frac{d-3}{4\pi r_1} \left[1 - \left(\frac{r_2}{r_1} \right)^{d-3} \right]$$

where $r_1 > r_2$ is the outer horizon. Deduce that the extremal Reissner–Nordström black hole has zero temperature in all spacetime dimensions. For $d = 4$ show that the temperature of the non-extremal black hole can be written as

$$T = \frac{m r_+ - q^2}{2\pi r_+^2}.$$

6.7.4 *Black hole entropy*

We now want to apply the connection between black holes and string theory (or its supergravity low energy form) to derive the entropy of a black hole. The problem is quite complicated so we can only outline the approach. The idea is to look at specific higher dimensional constructions that lead to black holes and count their degrees of freedom. We look at an example of a black hole in five dimensions that has three types of charge.

In conventional field theories charges reside on particles or 0-branes. String theory requires the existence of higher dimensional structures which support a generalisation of electric charge. Just as point charges are quantised in terms of integer multiples of the electron charge (or more fundamentally as integer multiples of quark charges), so too do the brane charges come in integer multiples. One can therefore think of multiple charged branes as coincident single branes. The five-dimensional picture is obtained from ten dimensions by compactifying the extra dimensions into small circles around which the branes become wrapped. The metric is

$$d\tau^2 = f(r)^{-2/3} dt^2 - f(r)^{1/3}(dr^2 + r^2 d\Omega_3^2), \tag{6.27}$$

where

$$f(r) = \left[1 + \frac{r_1^2}{r^2}\right]\left[1 + \frac{r_2^2}{r^2}\right]\left[1 + \frac{r_3^2}{r^2}\right].$$

The horizon is at $r = 0$ and its area is (see problem 6.10)

$$A = 2\pi^2 r_1 r_2 r_3.$$

If $r_1 = r_2 = r_3$ the metric reduces to that of the Reissner–Nordström solution for an extremal charged black hole in five dimensions. It therefore represents an extremal black hole with a horizon temperature of zero (see problem 6.9). The mass of the hole is

$$m = m_1 + m_2 + m_3 \quad \text{where } m_i = \frac{\pi r_i^2}{4G_5}.$$

The metric arises through Einstein's equations from charges carried by strings (D1 branes) and by D5 branes. It turns out that the charges are related to the masses (recall this is an extremal black hole) by

$$M_1 \propto Q_1 R; \quad M_2 \propto Q_5 R \quad M_3 = n/R,$$

where n is an integer. The charge Q_1 is that of a D1 brane on a circle, S; Q_5 is the charge of a 5 brane on a torus containing the circle S; the integer n gives the momentum carried by the branes round the circle. (We saw in Section 6.5.1 that momenta and charge were related.) The entropy can be obtained from the horizon area; in terms of the charges, it turns out to be

$$S = 2\pi k \sqrt{Q_1 Q_5 n}.$$

Now we count the ways of arranging for the strings to carry the momentum. The dominant term turns out to be the number of ways in which a long string wound $Q_1 Q_5$ times round the circle can have an excitation energy n/R. We can do no more than quote the result: for large $Q_1 Q_5 n$ the dominant term agrees with the Hawking entropy:

$$S = k \ln \Omega \sim 2\pi k \sqrt{Q_1 Q_5 n}.$$

Problem 6.10. *The area of the 3-sphere*

$$ds^2 = R^2 d\Omega_3^2 = R^2 (d\theta^2 + \sin^2 \theta d\phi^2 + \sin^2 \theta \sin^2 \phi d\psi^2)$$

is $2\pi^2 R^3$. Show that the area of the horizon at $r = 0$ in the metric 6.27 is $2\pi^2 r_1 r_2 r_3$.

The calculation can be performed for many other black holes that can be obtained from string configurations and agreement with the Hawking entropy is always found. Furthermore, when the calculation is done accurately the agreement is exact, including the universal numerical factor of $1/4$ between the entropy and the horizon area in Planck units. Where does the universality of this relation come from? In other words, how could we do the calculation once for all, rather than for each black hole separately? We consider one suggestion.

We begin by generalising conserved quantities from energy, to quantities constructed from the full energy momentum tensor. Such quantities are defined by Virasoro operators L_m, with m an integer, $(m = -\infty, \infty)$. By way of illustration we consider the BTZ black hole: the Hamiltonian is $H = 2L_0/l$ and (since the Hamiltonian is the energy, so $H = M$ for the black hole ground state) $L_0 = Ml/2$.

Next we consider the trace of the quantised energy momentum tensor: this is called the conformal anomaly. In the case of the BTZ black hole it is $T = -c \, ^{(2)}R/24\pi$ where c, called the central charge, is $c = 3l/2G$ and $^{(2)}R$ is the Ricci scalar in two dimensions.

Then (a simplified form of) the Cardy formula gives the number of states for a system in a state with eigenvalue Δ of L_0 as

$$\Omega(\Delta) \sim exp\left\{ 4\pi\sqrt{\frac{c\Delta}{6}} \right\} \rho(0).$$

For the BTZ black hole $\Delta = L_0 = Ml/2$ and the total degeneracy is

$$\ln \Omega \sim 4\pi\sqrt{cL_0/6} = 2\pi\sqrt{\frac{l^2 M}{2G}}$$

which is the Hawking entropy, equation (6.3). The hope is that the argument (or something similar) can be formulated in such a way as to apply to any black hole metric (Carlip, 2005, 2009).

6.7.5 *The information paradox and fuzzyballs*

We now return to the information paradox (Section 4.7.2). A key issue is whether we need to modify quantum theory radically to allow non-unitary evolution (non-conservation of probability) at a fundamental level. From the point of view of the AdS/CFT correspondence, quantum bulk properties, including quantum gravity, are determined by a normal quantum field theory on the spacetime boundary, which must evolve states in a unitary manner. So where this correspondence holds we know that quantum gravity must be a unitary theory. Since the correspondence includes black hole solutions, it resolves the information paradox in principle: the quantum evolution of a black hole must be unitary.

However, this argument only partially resolves the paradox: it does not tell us how to describe the evaporation of a black hole in terms of tracking the entropy. String theory provides a mechanism by which the radiation from a black hole is a pure state just like any other radiation process. The proposed picture involves the fact that in string theory, the quantisation of gravity can have important effects at a scale different from the Planck length l_P. A black hole may involve not one but many, say N, strings. Then the length at which quantum effects come into play will be Nl_P which can be of order of the horizon size.

This leads to the fuzzyball picture of black holes. The black hole mass is carried by strings spread out over the putative horizon. These radiate without the information ever being lost behind a horizon, since a true horizon never forms. The evaporation of stringy matter, called fuzzyballs, is then the same as the radiation from a conventional hot body.

6.8 Observations of higher dimensions

Finally we consider the prospects for verifying the existence of higher dimensions by observation. If the higher dimensions are compact, we have argued that in principle they should become visible to probes with high enough energies. It turns out that the cross-section for the formation of black holes in particle collisions is highly dependent on the spacetime dimension.

To see why this is, consider Newton's law of gravitation. In four spacetime dimensions the usual inverse squre law is

$$F = G\frac{m_1 m_2}{r^2}.$$

In d-dimensions this becomes

$$F = G_d\frac{m_1 m_2}{r^{d-2}}.$$

In order to maintain dimensional consistency the Newtonian constants G and G_d must be related by

$$G_d = l_*^{d-4}G$$

where l_* is some length scale. If l_* is the Planck length l_P then the strength of gravity is the same whatever the dimensionality of spacetime. However, in string theories the string scale l_* is not tied to the Planck scale. If $l_* \gg l_P$ and $d > 4$ then gravity is much stronger in the bulk spacetime than in our four-dimensional slice.

The effect is quite dramatic in terms of the energy requirements and production cross-section for mini black holes in particle collisions. The mass corresponding to the length scale l_* is $M_* = h/c^2 l_*$. This is the lower limit on energy required to produce black holes in particle collisions (to within a factor of 10). Various observations set a lower limits on M_* of around a few TeV. The Large Hadron Collider (LHC) can achieve a centre-of-mass energy of 14 TeV, so the production of mini black holes is energetically possible under favourable assumptions. In fact, if $d = 10$, $M_* = 1$ TeV the LHC could produce black holes of mass 5 TeV at the rate of one per second. These black holes would decay immediately by Hawking radiation, so their production would be verified by the detection of the appropriate spectrum of particles of various spins.

Chapter 7

Astrophysical Black Holes

7.1 Introduction

In the preceding chapters we have explored the properties of two highly symmetric vacuum solutions of the field equations of Einstein's general theory of relativity in four spacetime dimensions, the Schwarzschild and Kerr black holes. For many years it was expected that these black holes would turn out to be artefacts of high symmetry and that real stars, being neither exactly spherically nor axially symmetric, would not collapse to black holes. In that case black holes would be artificial mathematical constructs. On the other hand, if real non-symmetric stars do collapse to black holes one might expect there to be lots of differently non-symmetric black holes for them to collapse to. But we already know that, as far as electrically neutral vacuum solutions in asymptotically flat spacetimes are concerned, there are only the two black hole solutions of general relativity in four dimensions, a property that, as we noted, is described by saying that black holes have no hair. The situation seems somewhat surprising.

The resolution of the paradox is even more surprising! It is that in relativity, in contrast to Newtonian gravity, a collapsing body grows increasingly symmetric as the collapse proceeds. The difference arises because in general relativity an asymmetric body radiates away energy in the form of gravitational waves (wave-like oscillations of the metric coefficients $g_{\mu\nu}$) and in so doing radiates away its asymmetry. The problem can be treated by looking at the behaviour of small non-axially symmetric perturbations of the gravitational field during the collapse process (Price, 1972). The result is that multipole moments of the matter distribution are radiated away leaving, in general in the exterior region, a Kerr black hole. More generally one can use topological arguments to show that a black hole must form, and hence that it must be a Kerr or Schwarzschild black hole (Hawking and Ellis, 1973).

According to general relativity therefore, black holes are described exactly by the Kerr metric (including the Schwarzschild metric as a special case). We do not expect charged black holes to exist in nature, as such an object would rapidly attract opposite charges and be neutralised. So there are two questions we need to answer: do black holes exist, and if so are they really Kerr black holes? In this chapter we review the observational evidence.

The earliest evidence for the existence of black holes has come from the study of binary star systems. Within the Galaxy a number of binary systems are known that consist of a visible star and a compact dark companion in close orbit about each other. In some of these systems the orbital parameters of the two bodies can be determined and a lower limit to the mass of the dark companion obtained. When this mass is found to be greater than the limiting mass for a stable neutron star $(2\text{--}3\,M_\odot)$ then the dark companion becomes a strong candidate to be a black hole: we know of no physical mechanism that can balance the inward pull of gravity in a compact 'dead' star once its mass exceeds $3M_\odot$. Several dozen such candidates have been found. However, we can be certain that black holes really do exist only when we have demonstrated the existence of an event horizon, since this is the unique feature that defines a black hole and sets it apart from all other astronomical objects.

The discovery of stellar mass black hole candidates does not exhaust the possibilities. The study of active galaxies and also of the central region of our own Galaxy provides strong evidence for the existence of black holes with masses in the range $10^6\text{--}10^{10}\ M_\odot$, the so called supermassive black holes. It is now believed that such a massive black hole resides at the centre of nearly every galaxy, including our own.

More recent evidence suggests that there may also be black holes with masses intermediate between stellar mass black holes and supermassive black holes. These objects have been detected within neighbouring galaxies and in two cases at the centres of globular clusters, the dense star systems that orbit about the Milky Way.

One other possibility is that the universe could contain a population of low mass black holes, the so-called mini black holes. The evidence for these is so far inconclusive.

Overall, we shall see that the accumulated evidence for the existence of black holes is now very compelling.

Theories of gravity other than general relativity also predict the existence of black holes. The question whether the black hole candidates are Kerr black holes therefore constitutes a test of general relativity. This is a more difficult question to answer but, as we shall see, the X-ray emission of black hole binary systems does appear to accord with accretion by a Kerr black hole.

7.2 Stellar mass black holes

7.2.1 *Formation*

The theory of stellar evolution predicts that a star with a main-sequence mass exceeding about eight solar masses will end its life as a supernova, leaving behind a neutron star. In the cores of such massive stars nuclear burning goes all the way up to ^{56}Fe or ^{56}Ni (depending on the n:p ratio). These nuclei lie at the maximum of the nuclear binding energy curve, so energy-releasing reactions cease once this stage is reached. Once the mass of the iron core of such a star has built up to $1.4M_\odot$, the Chandrasekhar mass, the core will undergo gravitational collapse to form a neutron star. The gravitational energy released by the collapse appears initially as kinetic energy of the in-falling matter. Once nuclear densities are reached the collapse is halted, by nuclear forces and neutron degeneracy pressure, and the subsequent rebound launches a shock wave outwards through the core and into the envelope of the star. The shock wave heats the core to temperatures of the order of 10^{11} K but retains enough energy to blow off the envelope of the star. The result is a supernova. The bulk of the energy generated by the collapse is rapidly radiated from the very hot core in a large pulse of neutrinos that may also play a role in blowing off the envelope of the star.

In 1987 two neutrino detectors, one in Japan and one in the USA, simultaneously registered a short burst of about 12 neutrinos. A few days later supernova SN 1987A was observed in the large Magellanic cloud. The time delay and the numbers of neutrinos detected were consistent with the source of the neutrinos being SN 1987A at a distance of 50 kpc.

Although the equation of state of neutron matter is currently unknown, we can say that the upper limit to the mass of a neutron star cannot be greater than about $3M_\odot$. This estimate comes from setting the speed of sound in neutron matter equal to the speed of light; this condition leads to the stiffest (least compressible) equation of state consistent with causality. In fact the upper limit to the mass of a neutron star may not exceed $2M_\odot$ (Cottam *et al.*, 2002).

It seems likely that the scenario outlined above for the formation of a neutron star and the accompanying supernova applies to the collapse of a star with a main sequence mass that is less than about $30M_\odot$. It now appears that the collapse of a more massive star forms a black hole rather than a neutron star. The signature of such an event is an exceptionally energetic supernova, or hypernova as it has been dubbed, accompanied by a short very intense gamma-ray burst (Hjorth, 2003). These gamma ray bursts were first detected more than 30 years ago and have for some time been the subject of intense study. It is only recently that this connection between the long duration gamma-ray bursts, those lasting more than a few seconds, and supernovae has been established.

Another likely process for the formation of a black hole is the merger of the two members of a double-neutron-star binary system. The best known example of such a system is the binary pulsar discovered by Hulse and Taylor in 1974 (Taylor, 1994). Observations of this system have established that the two stars are slowly spiralling in towards each other. The system is losing energy and angular momentum, in the form of gravitational radiation, at a rate that is in agreement with the predictions of general relativity. It is estimated that the two stars will merge in about 3.2×10^8 years from now. As the total mass of the system is $2.83 M_\odot$ the final product of the merger depends on the equation of state of neutron matter, which is currently unknown. If the equation of state is soft, that is if the material is more easily compressed, the result will be a black hole.

It is believed that the short duration gamma ray bursts, those having a mean duration of about 0.2 seconds, may be the result of such neutron star mergers. It is anticipated that the burst of gravitational radiation emitted in the final stages of a merger will be measurable with gravitational wave detectors such as VIRGO in Italy, LIGO in the USA and GEO in Germany.

Recently another neutron star binary pulsar has been discovered (Burgay *et al.*, 2003) and found to contain two pulsars. This system has an orbital period of only 2.4 hours, a third of that of the Hulse–Taylor system, and an estimated time before the two stars merge of 8.5×10^7 years. The discovery of this new system has increased the expected rate of mergers. The detection rate by existing gravity wave detectors could be as high as once every 1–2 years.

7.2.2 *Finding stellar mass black holes*

An isolated black hole is completely dark so it can be detected only through its gravity. One way that this can be done is through the gravitational lensing of a background star when a dark object passes in front of it. The effect of a gravitational lens is to increase the intensity of light coming from the background star. The intensity increases smoothly to a maximum and then dies back down giving rise to a symmetrical light curve. If the distance to the lens can be obtained its mass can be calculated from the light curve. If the mass of the lens is greater than $3M_\odot$ then the object is a black hole candidate. Gravitational lensing has been observed, but to date none of the lens stars appears to be a black hole candidate.

All the known stellar mass black hole candidates are located in X-ray emitting binary star systems that consist of a star and a compact companion object in a close orbit about each other. In close binaries tidal forces tend to circularise the orbits. If the orbital parameters of the binary system can be obtained, a lower limit to the mass of the compact object can be estimated as follows. The time variation of the Doppler shift of absorption lines in the stellar spectrum yields the orbital period P of the system and also the amplitude of the line of sight velocity V_s of the star. These quantities determine the mass function

$$F(M) \equiv \frac{PV_s^3}{2\pi G} = \frac{\sin^3 \phi M_d}{(1 + M_s/M_d)^2},$$

where M_s is the mass of the star, M_d is the mass of the dark companion and ϕ is the angle between the line of sight and the normal to the orbital plane. Notice that $F(M)$ is a lower limit to M_d so if $F(M) > 3M_\odot$ we have a black hole candidate.

The first such black hole candidate to be discovered by this method was Cygnus X-1, the close binary companion of a blue giant star. For some time the mass of this blue giant companion was uncertain, because of a lack of an accurate distance measurement to the system, so various published values of the mass of Cygnus X-1 ranged from 5 to 20 solar masses. In order to get an accurate distance the very small parallax of the system has to be measured. Recently, the measurement of this parallax has become possible using the Very Long Baseline Array, a string of radio antennae stretching 8000 km from Mauna Kea in Hawaii to St Croix in the Virgin Isles. The distance comes out to be 1.86 kpc. Together with a better estimate of the viewing angle, ϕ, this gives a mass of $14.8 \pm 1M_\odot$ (Orosz *et al.*, 2011). This high value for the mass rules out the possibility of a neutron star and makes Cygnus X-1 one of the strongest black hole candidates.

A determination of the spin of Cygnus X-1 has been made by locating the inner edge of the accretion disc which is feeding in gas from the giant companion. This fixes the position of the marginally stable circular orbit.

the spin parameter a comes out to be $\approx 0.95m$, very close to the maximum (Lijun Gou *et al.*, 2011).

Since the discovery of Cygnus X-1 dozens of other black hole candidates have been found in binary systems in our Galaxy. Recently the first example of an eclipsing X-ray binary system, has been discovered in the nearby galaxy M 33 (Orosz *et al.*, 2007). The importance of this discovery is that the angle ϕ between the line of sight and the normal to the orbital plane is known to an accuracy of 1.5%. The mass of the companion is determined to be $70 \pm 7 M_\odot$ giving a mass for the dark companion of $15.65 \pm 1.45 M_\odot$. This mass is well above the maximum mass for a neutron star making this object the strongest candidate to be a black hole yet discovered.

Now $F(M) \gtrsim 3 M_\odot$ is a necessary condition for an object to be a black hole, but on its own it is not sufficient. Definitive proof requires evidence for an event horizon since this distinguishes a black hole from an object with a surface. A systematic comparison of neutron star binaries with candidate black hole binaries has found a difference between the two types of binary that is ascribable to the presence of a solid surface in the neutron star case and its absence in the black hole candidate case.

In both types of binary the two stars are close enough together for material from the visible star to be pulled off and captured by the companion. Because the material has angular momentum it tends to orbit about the compact object forming a pancake shaped accretion disc. A packet of gas in the disc at radius r will move in a nearly Keplerian orbit with velocity $v \propto r^{-1/2}$. As the gas making up the disc has turbulent viscosity neighbouring annuli rubbing against each other will give rise to energy dissipation. This causes the gas to spiral slowly inwards to smaller radii and heats the gas in the inner part of the disc to temperatures high enough for it to emit X-rays. If the compact object is a neutron star the gas will eventually hit its surface and accumulate there. The kinetic energy of the in-falling gas is converted to heat as it strikes the surface of the neutron star and gravity on the stellar surface compresses the gas to the point where it undergoes thermonuclear burning. If the burning is unstable the result is a violent explosive outburst of X-ray emission. Such bursts are observed from neutron star binaries. We would not expect to see such an outburst from an accreting black hole as it has no surface. And no instance of such a burst has been seen from a black hole candidate (McClintock *et al.*, 2003). This is evidence that the candidate black holes have an event horizon.

Problem 7.1. *Derive the expression quoted above for the mass function of a binary star system.*

7.2.3 *The black hole at the centre of the Galaxy*

Lynden-Bell and Rees suggested in 1971 that there might be a massive black hole at the centre of the Galaxy. In this section we look at the current evidence for this.

In 1974 a strong compact radio source was detected in the direction of the Galactic centre and the discovery was confirmed by further observations the following year. This source was named Sagittarius A* in 1982 by Brown.

The central few parsecs of the Galaxy also contain a cluster of stars having a number density of about 10^6 pc^{-3}. The proper motions of individual stars in this central region have been followed over a period of several years and show that the stars are moving about the position of Sagittarius A*. Thus Sagittarius A* is located at the dynamic centre of the Galaxy. In particular more than one complete orbit of a star named SO-2 has been plotted out. The orbit is highly elliptical with a major axis of 5.5 light days and a period of 15.56 years. The point of closest approach is only 17 light hours from Sag A*. This means that there are $4 \times 10^6 M_\odot$ confined within a volume comparable to that of the Solar System. A cluster of dark stars confined within such a volume can be ruled out: it would have a density of order 10^{15} pc^{-3}, too high to exist as a stable system. The only known explanation is that there is a black hole at or near to the position of Sag A*. Over a single orbit of SO-2 no departure from Keplerian motion has been detected: a number of orbits will be needed before relativisitic precession can be measured. Recently another, fainter star, designated SO-102, has been detected on a closer orbit with a period of about 11.5 years (Meyer *et al.*, 2012). This star promises to provide an even better probe of the gravitational field of Sag A*.

There is some evidence that Sag A* may have a significant angular momentum. Observations of Sag A* in infrared radiation have detected flares which exhibit variability with a period of about 17 minutes (Genzel *et al.*, 2003). The flares also have rapid rise and decay times, which suggests that they are generated in the innermost zone of the accretion disc. If this is the case then the 17 minute period could be the orbital period of matter close to the last stable orbit of an accretion disc around the black hole.

As we can see from equation (3.6) in Section 3.13.2 the orbital period depends on the spin of the hole mainly through its effect on the radius of the last stable orbit. For a Schwarzschild black hole of mass $4.0 \times 10^6 M_\odot$ the period of matter in the last stable orbit is about 28 minutes. So a period of 17 minutes is too short for a slowly spinning black hole of the mass of Sag A*. As we can see from Fig. 3.2 in Chapter 3, the radial coordinate of the last stable orbit of a particle in a prograde orbit decreases with increasing angular momentum parameter a. In fact a 17 minute period is obtained for a prograde orbit when the black hole has $a = 0.52m$ (problem 3.19). Note from Fig. 3.2 that the radial coordinate of retrograde orbits increases with increasing a so a 17 minute period rules out retrograde orbits in this case.

The calculations of the orbital periods quoted above assume that the orbits lie in the equatorial plane. This is justified for the inner region of an accretion disc, since Lense–Thirring precession and viscous torques will force an out-of-plane orbit into the equatorial plane (Bardeen and Petersen, 1975).

Problem 7.2. *Analysis of the X-ray spectrum from the accretion disc around the black hole at the centre of the galaxy NGC 1365 indicates that its inner edge (i.e. the position of the marginally stable orbit) lies at $r = 1.2m$.*
(a) What value of the spin parameter is implied by this result?
(b) What is the binding energy of a particle in the marginally stable orbit at $r = 1.2m$?

7.3 Supermassive black holes in other galaxies

Observations of the centres of other galaxies cannot be as detailed as those of the Milky Way but again it is found that material near their centres is orbiting rapidly around a central mass concentration. In addition some galaxies emit large amounts of energy from a very small volume at their centres. Quasars are an extreme example of these active galaxies. The luminosity from the centre of a quasar can exceed the output from the rest of the galaxy a thousand fold, yet this energy is coming from a volume not much larger than that of the Solar System. The only mechanism that can reasonably explain such a high luminosity from such a small volume is the accretion of matter by a black hole at a rate of up to several solar masses per year. Recall, from Chapter 3, that a lump of matter that is accreted from the last stable orbit of an extreme Kerr black hole must have lost about 30% of its rest energy in the process of spiralling in. On the other hand, nuclear burning in stars can release no more than 0.7% of the rest energy of a star. It is hard to envisage how enough stars to produce the observed luminosity could exist within such a small volume as a stable system. On the other hand, a black hole explains the luminosities and other features of active galaxies convincingly.

The Chandra X-ray observatory satellite has found two supermassive black holes in orbit about each other in the galaxy NGC 6240. The exceptionally high rate of star formation in this galaxy suggests that it formed recently from the merger of two smaller galaxies. If each contained a central black hole this would explain the presence of two black holes near the centre of the merged galaxy.

The process by which supermassive black holes form at the centres of galaxies is not yet understood, but at the time of writing there are some clues (see Section 7.3.1). They are thought to form at an early stage in the history of a galaxy and to power the quasar activity observed in young galaxies (Barkana and Loeb, 2003).

A further intriguing possibility that is close to realisation is, in a sense, actually to 'see' the event horizon of Sagittarius A*. The resolution of very long baseline radio interferometers (VLBI) is approaching the scale of the event horizon of Sagittarius A* (Doeleman *et al.*, 2008). Calculations have been carried out that simulate the appearance of the emitting gas surrounding Sagittarius A* (Melia and Falcke, 2001). These show that under high resolution VLBI a central shadow of diameter $\sim 10m$, for a black hole of mass m, would be visible in the radio image of the Galactic centre. The detection of such a shadow would constitute direct evidence for the existence of an event horizon.

7.3.1 *Intermediate mass black holes*

Over the past few years a number of highly luminous compact X-ray sources have been detected in nearby galaxies. These sources emit 10–100 times as much X-ray power as stellar mass black holes in X-ray binaries. At present their nature is unclear but they may be accreting black holes with masses of the order of 100–$1000 M_{\odot}$. In particular they are often found in galaxies where an episode of rapid star formation is underway – so called star-burst galaxies. This suggests that they may be formed by the merging of stellar mass black holes produced in large numbers in the star burst. It has been suggested that intermediate mass black holes could sink to the core of a galaxy where they would merge to form a supermassive black hole. These intermediate mass black holes could therefore be the key to understanding how the supermassive black holes form.

There is also evidence for black holes at the centres of globular clusters. Measurements with the Hubble space telescope of the velocities of stars in the central regions of two globular clusters point to the presence of black holes there. The putative black hole at the centre of the globular cluster M15, which is in orbit about our Galaxy, is estimated to have a mass of about 4000 $M_\odot$ (Van Der Marel *et al.*, 2002). The very large globular cluster G1 attached to the Andromeda galaxy appears to harbour a central black hole of about $2 \times 10^4 M_\odot$ (Gebhardt *et al.*, 2002). In both of these cases the relationship between the black hole mass and the cluster mass bears a similar relationship to that between the mass of a galactic black hole and the mass of its host galaxy.

7.3.2 *Mini black holes*

There is no lower limit to the mass of a black hole so in principle black holes with masses $\ll 1 M_\odot$ can exist. However the lower the mass of the black hole the greater the density to which matter has to be compressed to create it. This can be seen from equation (1.1) which states that $\rho \propto M^{-2}$. Take for example a black hole having a mass of 10^{12} kg. Its density must be of the order of 10^{57} kg m^{-3} (10^{39} times the density of a neutron star). Only in the very early universe, when ultra-high densities prevailed, could suitable conditions have existed for the formation of such small black holes.

From Chapter 4 we recall that black holes emit Hawking radiation and have a lifetime $t \propto M^3$. Thus, any primordial black holes that exist today must have had an initial mass of at least 10^{12} kg, since any lighter black holes would have evaporated away before the present time.

How would an evaporating black hole nearing its final stage manifest itself? As its mass decreases its temperature increases according to the Hawking relation equation (4.29)

$$T_h = \frac{\hbar c^3}{16\pi k G M}.$$

The criterion for emission is that as soon as the temperature exceeds the rest mass of a particle species that species starts to be emitted by the black hole. At lower temperatures photons, neutrino antineutrino pairs and gravitons would be emitted. As the mass of the hole declines and the temperature increases then to these particles would be added successively electron-positron pairs followed by the heavier lepton anti-lepton pairs and then quark anti-quark pairs and beyond these, if they exist, the supersymmetric particles. At some point the temperature reaches the stage where quantum gravity effects come into play. As we do not yet have a theory of quantum gravity the final end point of the evaporation process is not known. The end point could be a stable object with a mass close to the

Planck mass, in which case such relics would contribute to the dark matter that makes up most of the mass of a galaxy.

What is the observational evidence for the existence of primordial black holes? The burst of gamma rays and other particles emitted in the final stage of evaporation of a single black hole are unlikely to be detected, but the gamma radiation from the population of such objects would contribute to the cosmic gamma ray background. Measurements of the gamma ray background give an upper limit to the present mass density of primordial black holes which is less than $10^{-8}\Omega$, where Ω is the cosmological density parameter having a value close to 1.

On the other hand, the quarks emitted in the evaporation process would give rise to jets of hadrons that decay ultimately to photons, leptons, protons and antiprotons. These jets are similar to those produced in collisions of high-energy electrons and positrons at, for example, the LEP particle accelerator. The antiprotons could account for the unexpectedly large fraction of antiprotons that are detected in cosmic rays. Another anomaly is that the electron-positron cosmic ray background has a 50% positron content in the vicinity of 100 MeV. Such a high ratio is hard to explain without a contribution from primordial black holes. (MacGibbon and Carr, 1991).

However, we must conclude that at present the evidence for the existence of primordial black holes is inconclusive.

7.4 Further evidence for black hole spin

In Section 7.2.3 we cited some evidence for spin in Sagittarius A*. Here we look at some evidence for the spin of stellar mass black holes in binary systems.

We saw in Chapter 3 that the angular momentum of a black hole strongly affects the orbits of particles and light rays close to the event horizon. Thus, for example, the spin of a black hole will influence the way in which matter from an accretion disc is captured. For an extreme Kerr black hole, in which the gas is orbiting in the same sense as the hole, the binding energy released by the gas as it spirals inwards towards the last stable orbit is about 30% of its rest energy. On the other hand, for gas with angular momentum in the opposite sense to that of an extreme Kerr black hole the last stable orbit is at $r_{ls} = 9m$, and the binding energy liberated up to this point is only about 3%. For a non-rotating black hole $r_{ls} = 6m$ and the binding energy released is about 6%.

Many black hole candidates in binary systems emit an ultra-soft X-ray component with a blackbody (colour) temperature in the range 0.5 to 2 KeV. Zhang and Cui (1998) have suggested that the strength of this soft X-ray component is directly related to the spin of the black hole. They base their conclusion on the calculation of the X-ray flux from the surface of a thin accretion disc in the equatorial plane of a Kerr black hole. According to their calculations the temperature and luminosity of the soft X-rays are greatest for an extreme Kerr black hole rotating in the same sense as the accretion disc, prograde rotation, and least for the extreme retrograde case. Based on their soft X-ray spectra, black hole candidates can be assigned to three categories: extreme prograde, extreme retrograde and non-spinning or slowly spinning systems. However, two of the black hole candidates, one of which is notably Cygnus X-1, appear to switch periodically from a retrograde system to a prograde system. Numerical simulations show that the switching of accretion flow is not as improbable as it might at first seem. Unfortunately the physics of accretion discs is not yet understood well enough for accurate models of the accretion process to be constructed.

Jets of ionised plasma travelling at relativistic velocities are observed from both stellar mass and supermassive black hole systems. The idea is that a pair of back-to-back jets are directed along the spin axis, or symmetry axis, of the black hole. Often only a single jet is visible. Since the jets are travelling at relativistic velocities, the idea is that aberration will enhance the luminosity of a jet with a component of motion in our direction, and will diminish the luminosity of the other jet. It is thought that the power source for jet production is the rotational energy of the black hole. We saw in Chapter 3 that orbits having negative angular momentum and negative energy exist within the ergosphere region of a spinning black hole, and that these orbits can be used to extract energy and angular momentum from a black hole using the Penrose process. The Penrose process is however not a practical mechanism for energy extraction (Section 3.15.4). It seems most likely that some other process powers astrophysical jets, probably involving the coupling of the rotation of the hole to exterior matter by magnetic fields as in the Blandford–Znajek mechanism. (For a non-technical account see Begelman and Rees, 1998.)

Current work on the measurement of spin involves the observation of lines of ionised iron in the X-ray region in the emission spectrum of active galactic nuclei. These lines are formed in gas close enough to the black hole for their spectroscopic profiles to be affected by general relativistic redshifts in addition to the normal Doppler shifts that arise from the motion of the gas. In principle the black hole parameters (mass and angular momentum) and the orientation of the accretion disc can be determined from these line profiles. In practice, the profiles depend on models of the emission region, and there may also be other contributions to the line broadening from scattered radiation. In addition the observations have limited resolution. So this work has some way to go.

In future X-ray polarisation measurements of the line emission from matter close to the black hole will also yield the parameters of the hole and act as a test of relativity.

7.5 Colliding black holes

Just as in a neutron star binary system the members of a black hole binary will spiral in towards each other and finally merge to form a single black hole. In the later stages of the in-spiralling, gravitational radiation becomes the principal mechanism by which energy and angular momentum are carried away from the system. Of the order of 4% of the mass of the binary is converted into gravitational waves in such a merger; this corresponds to a luminosity of $\sim 10^{23} L_\odot$. Black hole mergers are prime candidates to be detected by the gravitational wave detectors currently under development.

In order to recognise a black hole merger, we must turn to numerical computations in general relativity to provide the wave-form of the gravitational waves emitted by a binary system as it spirals inwards up to the point of merger. This is a formidably difficult task which has taken about forty years to accomplish. Until recently, calculations became unstable even before a single orbit had been completed. But these instabilities have now been eradicated by new numerical techniques and it is possible to follow the evolution of the binary system, and obtain its gravitational waveform, to the point of merger and thereafter through the 'ringdown' phase where the non axi-symmetric oscillations of the resulting black hole are radiated away to leave a Kerr black hole. (For an historical account see Centrella *et al.*, 2010.) The calculation can be performed at the time of writing for ratios of masses of the binary pair up to about 9 and for spin parameters up to $a \sim 0.9$. Comparison of these calculations with observation would provide a test of general relativity in the strong field regime.

7.6 Conclusions

The existence of stellar mass black holes, and of supermassive black holes at the centres of most, if not all, galaxies is supported by an abundance of evidence. There is little doubt that such objects exist as described by the general theory of relativity. Only recently has it been possible to detect intermediate mass black holes, and the evidence for them is slighter but growing. It is likely that these three categories of black holes are related. The current picture is that in the centres of galaxies, and of globular clusters, stellar mass black holes merge and grow to supermassive black holes and intermediate mass black holes respectively. On the other hand, mini black holes are not the products of stellar evolution. Their existence would depend on the suitability of conditions in the early universe for their formation, so their existence is of great interest for cosmology.

Solutions to Problems

Problem 1.1. The Galilean transformation for infinitesimals is $dx = dx' + v\,dt'$, $dy = dy'$, $dz = dz'$ and $dt = dt'$. Substituting this into equation (1.7) $ds^2 = c^2dt^2 - dx^2 - dy^2 - dz^2$ gives $ds^2 = c^2dt'^2 - dx'^2 - dy'^2 - dz'^2 - 2v\,dx'\,dt' - v^2dt'^2 \neq ds'^2$. A similar result follows for the sum of squares. In general the reason why we cannot find an invariant under the Galilean transformation is that the time component is fixed, so changes in the x component cannot cancel changes in the time component.

Problem 1.2. Multiplying $v'^\mu = \frac{\partial x'^\mu}{\partial x^\nu}v^\nu$ by $\frac{\partial x^\rho}{\partial x'^\mu}$ gives $\frac{\partial x^\rho}{\partial x'^\mu}v'^\mu = \frac{\partial x^\rho}{\partial x'^\mu}\frac{\partial x'^\mu}{\partial x^\nu}v^\nu = \delta^\rho_\nu v^\nu = v^\rho$. Change the free index ρ to μ and the dummy index μ to ν and the result follows. A similar procedure yields the inverse transformation of the covariant components.

Problem 1.3. The scalar product $v^\mu v_\mu = \frac{\partial x^\mu}{\partial x'^\nu}v'^\nu \frac{\partial x'^\rho}{\partial x^\mu}v'_\rho = \delta^\rho_\nu v'^\nu v'_\rho = v'^\nu v'_\nu$. Hence the scalar product is an invariant.

Problem 1.4.

$$\nabla_\nu(A^\mu B_\mu) = \partial_\nu(A^\mu B_\mu).$$

Differentiating gives

$$B_\mu \nabla_\nu A^\mu + A_\mu \nabla_\nu B^\mu = B_\mu \frac{\partial A^\mu}{\partial x^\nu} + A^\mu \frac{\partial B_\mu}{\partial x^\nu}.$$

Now

$$\nabla_\nu A^\mu = \frac{\partial A^\mu}{\partial x^\nu} + \Gamma^\mu_{\nu\rho}A^\rho.$$

So

$$A^\mu \nabla_\nu B_\mu = A^\mu \frac{\partial B_\mu}{\partial x^\nu} - B_\mu \Gamma^\mu_{\nu\rho}A^\rho.$$

Interchange the indices μ and ρ in the last term and cancelling the A^μ gives

$$\nabla_\nu B_\mu = \frac{\partial B_\mu}{\partial x^\nu} - \Gamma^\rho_{\nu\mu}B_\rho.$$

Problem 1.5. Let

$$L_V A^\mu = V^\nu \nabla_\nu A^\mu - A^\nu \nabla_\nu V^\mu$$

$$= V^\nu \left[\frac{\partial A^\mu}{\partial x^\nu} + \Gamma^\mu_{\nu\rho} A^\rho \right] - A^\nu \left[\frac{\partial V^\mu}{\partial x^\nu} + \Gamma^\mu_{\nu\rho} V^\rho \right]$$

$$= V^\nu \partial_\nu A^\mu - A^\nu \partial_\nu V^\mu + \Gamma^\mu_{\nu\rho} A^\rho V^\nu - \Gamma^\mu_{\nu\rho} A^\nu V^\rho.$$

The last two terms cancel, as the dummy indices can be relabelled to make them identical, so it is legitimate to replace the partial derivatives by covariant derivatives in the Lie derivative.

Problem 1.6. The tensor transformation law for a second rank contravariant tensor is

$$T^{\mu\nu} = \frac{\partial x^\mu}{\partial x'^\sigma} \frac{\partial x^\nu}{\partial x'^\rho} T'^{\sigma\rho}.$$

The co-variant derivative is

$$\nabla_\sigma T^{\mu\nu} = \frac{\partial T^{\mu\nu}}{\partial x^\sigma} + \Gamma^\mu_{\sigma\rho} T^{\rho\nu} + \Gamma^\nu_{\sigma\lambda} T^{\mu\lambda}.$$

The Lie derivative is

$$\pounds_V T^{\mu\nu} = V^\sigma \nabla_\sigma T^{\mu\nu} - T^{\sigma\nu} \nabla_\sigma V^\mu - T^{\mu\sigma} \nabla_\sigma V^\nu.$$

Problem 1.7. We have $g^{\mu\nu} g_{\mu\nu} = 4$. Therefore

$$g^{\mu\nu}\left(R_{\mu\nu} - \frac{1}{2} g_{\mu\nu} R\right) = R - 2R = -R = g^{\mu\nu} T_{\mu\nu} = 0. \tag{1}$$

Problem 1.8. We have from problem 1.6

$$\pounds_K g^{\mu\nu} = k^\alpha \nabla_\alpha g^{\mu\nu} - g^{\alpha\nu} \nabla_\alpha k^\mu - g^{\mu\alpha} \nabla_\alpha k^\nu$$

$$= 0 - (\nabla^\nu k^\mu + \nabla^\mu k^\nu),$$

from which the result follows.

Problem 2.1. From the metric the proper circumference of the curve $r =$ constant in the plane $\theta = \pi/2$ at constant time t is $2\pi r$. At the event horizon $r = 2m$. So the proper circumference of a Schwarzschild black hole is $4\pi m$. From the metric the area of the surface $r = $ constant, $t = $ constant is

$$\text{Area} = r^2 \int_0^{2\pi} d\phi \int_{-\pi/2}^{\pi/2} \sin\theta d\theta.$$

The angular integral is the area of the unit sphere, i.e. 4π and $r = 2m$ so the black hole area is

$$A_h = 16\pi m^2.$$

Problem 2.2. The relationship between E and E_N for large r and small velocities is obtained by equating $(dr/dt)^2$ and $(dr/d\tau)^2$ from equations (2.11) and (2.18) respectively. After converting (2.18) to physical units we get

$$2E_N - \frac{L_N^2}{r^2} + \frac{2GM}{r} = -c^2 + \frac{E^2}{c^2} - \frac{L^2}{r^2} + \frac{2GM}{r} + \frac{2GML^2}{c^2 r^3}.$$

The final term is specific to relativistic gravity and is small at large r. Ignoring it we get $2E_N + c^2 = E^2/c^2$ as required.

Problem 2.3. The geodesic equations are

$$\frac{du^\mu}{d\tau} + \Gamma^\mu_{\sigma\rho} u^\sigma u^\rho = 0.$$

The $\mu = 1$ equation, in physical units, is

$$\frac{d^2 r}{d\tau^2} + \Gamma^1_{00} c^2 \left(\frac{dt}{d\tau}\right)^2 + \Gamma^1_{22}\left(\frac{d\theta}{d\tau}\right)^2 + \Gamma^1_{33}\left(\frac{d\phi}{d\tau}\right)^2 = 0.$$

With the Newtonian metric, $\Gamma^1_{00} = GM/c^2 r^2$, $\Gamma^1_{22} = -r$ and $\Gamma^1_{33} = -r\sin^2\theta$ and at low speeds, $dt \sim d\tau$. With these values the $\mu = 1$ equation becomes

$$\frac{d^2 r}{dt^2} = \frac{GM}{r^2} - r\left(\frac{d\phi}{dt}\right)^2$$

where the spherical symmetry confines the particle orbits to a plane and we have chosen the $\theta = \pi/2$ plane. This is the Newtonian expression for radial acceleration under gravity in polar coordinates.

Similarly the $\mu = 3$ equation becomes $r\frac{d^2\phi}{dt^2} + 2\frac{dr}{dt}\frac{d\phi}{dt} = 0$. This is the Newtonian expression for transverse acceleration in polar coordinates.

Problem 2.4. Expanding $1 = g_{\mu\nu} u^\mu u^\nu$ gives for radial motion

$$1 = g_{00}(u^0)^2 + g_{11}(u^1)^2.$$

Substituting for the metric coefficients with $u^0 = E/(1 - 2m/r)$ gives

$$1 = \left(1 - \frac{2m}{r}\right) \frac{E^2}{(1 - 2m/r)^2} - \frac{(u^1)^2}{(1 - 2m/r)}.$$

Finally, rearranging and taking square roots gives

$$u^1 = \pm \left[E^2 - \left(1 - \frac{2m}{r}\right)\right]^{1/2}.$$

Problem 2.5. The 4-velocity of an observer hovering at radius r is

$$u^\mu_H = \left(\left(1 - \frac{2m}{r}\right)^{-\frac{1}{2}}, 0, 0, 0\right).$$

The covariant components of the 4-velocity of a particle in radial free fall are

$$(u_\mu) = (E, u_1, 0, 0).$$

So the energy per unit mass of the falling particle in the frame of the hovering observer is given by

$$u^\mu_H u_\mu = \gamma = (1 - v^2)^{-\frac{1}{2}},$$

where v is the local velocity of the particle. Evaluating the scalar product in the Schwarzschild frame gives $E(1 - 2m/r)^{-\frac{1}{2}}$. But the scalar product is an invariant, so we can equate it to its value γ in a local inertial frame, hence

$$E(1 - 2m/r)^{-\frac{1}{2}} = (1 - v^2)^{-\frac{1}{2}}.$$

The velocity at r of a particle falling from infinity is just the velocity needed to project the particle back from r to infinity, i.e. the escape velocity; hence setting $E = 1$, the energy of a particle dropped from rest at infinity, gives

$$v_{\text{esc}} = \left(\frac{2m}{r}\right)^{\frac{1}{2}}.$$

At the horizon the escape velocity is the speed of light.

Problem 2.6. (a) Invert equation (2.26) to get L in terms of r and the result follows. (b) Substitute for L in equation (2.18) and set $dr/d\tau = 0$, the condition for a circular orbit.

Problem 2.7. From the geodesic equations (2.14) and (2.17)

$$\frac{dt}{d\tau} = \frac{E}{\left(1 - \frac{2m}{r}\right)} \tag{2}$$

and

$$\frac{d\phi}{d\tau} = \frac{L}{r^2} \tag{3}$$

where τ is the proper time of the orbiting particle. Substituting the expression for L obtained in problem 2.6 into (3) and integrating over an orbit gives

$$\tau = \frac{2\pi r^{3/2}}{m^{1/2}} \left(1 - \frac{3m}{r}\right)^{\frac{1}{2}}.$$

To obtain the period measured by a distant observer we divide (3) by (2):

$$\frac{d\phi}{dt} = \frac{d\phi/d\tau}{dt/d\tau} = \frac{L(1 - 2m/r)}{r^2 E}.$$

Now substitute for E and L the expressions obtained in problem 2.6 and integrate over one orbit to get

$$T = \frac{2\pi r^{\frac{3}{2}}}{m^{\frac{1}{2}}}.$$

Problem 2.8. The energy per unit mass of a particle in a circular orbit is

$$E = \left(1 - \frac{2m}{r}\right) \left(1 - \frac{3m}{r}\right)^{-\frac{1}{2}}.$$

Put $E = 1$, the energy of a particle at rest at infinity, and solve for r to get $r = 4m$. Inserting this value of r into $L = (mr)^{\frac{1}{2}} (1 - 3m/r)^{-\frac{1}{2}}$ gives $L = 4m$ thus $L/E = 4m$. This is an unstable orbit because it lies inside the last stable orbit at $r = 6m$. Note that in Newtonian physics a particle coming in from infinity must lose energy before it can be inserted into a circular orbit.

Problem 2.9. The velocity v of a particle in a circular orbit at radius r is related to the escape velocity v_{esc} from that radius by

$$v = \frac{v_{esc}}{\sqrt{2}(1 - 2m/r)^{1/2}}.$$

Clearly when $r = 4m$ then $v = v_{esc}$. For $r < 4m$ the value of the denominator falls with decreasing r and $v > v_{esc}$.

Problem 2.10. Since Velocity = (Proper Distance)/(Coordinate Time), the velocity of a particle in a circular orbit at radius r is

$$\frac{2\pi r}{2\pi (r^3/m)^{1/2}}.$$

At $r = 3m$ the particle orbit becomes a photon orbit and the velocity is $1/\sqrt{3}$ in geometrical units or $c/\sqrt{3}$ in physical units.

The period of the photon in its orbit acts as a local stationary clock at $r = 3m$: each orbit gives one tick of the clock. When viewed from infinity this period is lengthened by a factor $(1 - 2m/r)^{-1/2} = \sqrt{3}$, so the observed speed is decreased by a factor $1/\sqrt{3}$.

Problem 2.11. We evaluate $d\tau/m_0$ in a local inertial frame as we can use special relativity in such a frame. The infinitesimal timelike interval between a pair of events along the world line of a particle having velocity u in flat spacetime is $d\tau = dt(1 - u^2)^{1/2}$. The mass m_0 of the particle is related to its energy and momentum by $m_0 = (E^2 - p^2)^{1/2} = E(1 - u^2)^{1/2}$ as $p = Eu$: note that E here is the energy of the particle not the energy per unit mass. So $d\tau/m_0 = dt/E$. This expression is independent of the velocity of the particle so it also applies to photons. Hence $d\tau/m_0 \rightarrow dt/E_{ph}$ which is finite.

Problem 2.12. The 4-velocity of a hovering observer at an arbitrary radial coordinate r is $(u^\mu) = ((1 - 2m/r)^{-\frac{1}{2}}, 0, 0, 0)$ and the 4-momentum of a particle at r has covariant components $(p_\mu) = (E, p_1, p_2, p_3)$. To obtain the velocity of the particle with respect to the local hovering observer we evaluate the scalar product $u^\mu p_\mu = \gamma m_0$ in the local frame and in the Schwarzschild frame and equate the results. This gives $m_0 E(1 - 2m/r)^{-\frac{1}{2}} = m_0(1 - v^2)^{-\frac{1}{2}}$. Solving for v the velocity of the particle gives

$$v^2 = \frac{m_0^2 E^2 - (1 - 2m/r)m_0^2}{m_0^2 E^2}.$$

Now as $m_0 \rightarrow 0$, $m_0 E \rightarrow E_{ph}$ and the second term in the numerator goes to zero. Thus in the limit $v = 1$, the speed of light. An alternative way to do this is to cancel the m_0 and remember that E is the energy at infinity, so $E^2 = (1 - v_\infty^2)^{-1}$. Now $v^2 = 1 - (1 - v_\infty^2)(1 - 2m/r)$. Set v at infinity equal to the speed of light; then the local speed v is everywhere equal to the speed of light.

Problem 2.13. This calculation follows closely the calculation in Section 2.4.1. We evaluate the scalar product $u^\mu p_\mu = h\nu$, where u is the 4-velocity of the falling spaceship, p the 4-momentum of a photon coming from infinity and $h\nu$ is the energy of the photon in the spaceship frame. Now we evaluate the scalar product in the Schwarzschild frame which is the frame of an observer at infinity and equate it to $h\nu$. The components of the probe's 4-velocity in the Schwarzschild frame are $(u^\mu) = \left(\left(1 - \frac{2m}{r}\right)^{-1}, -\left(\frac{2m}{r}\right)^{1/2}, 0, 0\right)$ and the covariant components of momentum of the inwardly propagating photon are $(p_\mu) = \left(h\nu_\infty, h\nu_\infty\left(1 - \frac{2m}{r}\right)^{-1}, 0, 0\right)$. So finally we get

$$h\nu_\infty \left[1 + \left(\frac{2m}{r}\right)^{1/2}\right]^{-1} = h\nu.$$

The crew can continue to receive signals even when they are inside the horizon. Only as the singularity at $r = 0$ is approached does the signal become infinitely redshifted.

Problem 2.14. The conditions for a circular orbit are $\frac{dr}{d\lambda} = 0$ and $\frac{d^2r}{d\lambda^2} = 0$. Equation (2.30) gives

$$\left(\frac{dr}{d\lambda}\right)^2 = E_{ph}^2 - \frac{L_{ph}^2}{r^2}\left(1 - \frac{2m}{r}\right).$$

Differentiating we get

$$\frac{d^2r}{d\lambda^2} = \frac{2L_{ph}^2}{r^3}\left(1 - \frac{2m}{r}\right) - \frac{mL_{ph}^2}{r^4}.$$

With $L_{ph}/E_{ph} = 3\sqrt{3}m$ both of these expressions are zero at $r = 3m$. So the light ray is indeed in a circular orbit at this radius.

Problem 2.15. The apparent diameter of the hole is $3\sqrt{3}(GM/c^2) = 3.1 \times 10^{10}$ m. So at the Earth it subtends an angle $\sim 26\mu$as.

Problem 2.16. We start from

$$\left(\frac{dr}{d\phi}\right)^2 = \left(\frac{dr/d\tau}{d\phi/d\tau}\right)^2 = \frac{E^2 - (1 + L^2/r^2)(1 - 2m/r)}{L^2/r^4}.$$

Let $u = 1/r$ then

$$\frac{du}{d\phi} = -\frac{1}{r^2}\frac{dr}{d\phi}.$$

Changing to the new variable in the above expression gives

$$\left(\frac{du}{d\phi}\right)^2 = \frac{E^2}{L^2} - (1 + L^2 u^2)(1 - 2mu)/L^2,$$

which, on differentiating with respect to ϕ, gives

$$\frac{d^2 u}{d\phi^2} + u = \frac{m}{L^2} + 3mu^2.$$

In the Newtonian limit the term $3mu^2$ is zero so it represents the departure from Newtonian physics. Making the substitution $u = l^{-1}(1 + \varepsilon \cos \lambda\phi)$ and collecting together the cosine terms gives

$$\left(-\frac{\varepsilon\lambda^2}{l} + \frac{\varepsilon}{l} - \frac{6m\varepsilon}{l^2}\right)\cos\lambda\phi + \frac{1}{l} = \frac{m}{L^2} + \frac{3m}{l^2}.$$

For a solution to this equation the coefficient of the $\cos\lambda\phi$ term must vanish as there is no ϕ on the right-hand side; that is $\lambda = (1 - 6m/l)^{1/2} \sim 1 - 3m/l$ and $1/l = m/L^2 + 3m/l^2 = m/L^2$. In the Newtonian limit $\lambda = 1$ so here $3m/l$ is small for weak fields. In one orbit $\lambda\phi = 2\pi$ so $\phi = 2\pi/(1 - 3m/l) \approx 2\pi + 6\pi m/l$. Therefore in one orbit ϕ has turned through the extra angle $6\pi m/l = 6\pi m^2/L^2$. For a circular or near circular orbit $L^2 = mr/(1 - 3m/r) = mr$ for $r \gg m$. So the amount of precession per orbit is $\Delta\phi = 6\pi m/r$. For the planet Mercury, for example, this comes to 43 seconds of arc per century. For a light ray

$$\left(\frac{dr}{d\phi}\right)^2 = \frac{E_{ph} - (L^2/r^2)(1 - 2m/r)}{L^2/r^4}.$$

Setting $u = 1/r$ and differentiating gives

$$\frac{d^2 u}{d\phi^2} + u = 3mu^2.$$

In the absence of the term $3mu^2$ the solution to the equation is of the form $u = A\cos(\phi + \alpha)$, which is the equation of a straight line in polar coordinates. Finally setting $u = $ constant in the above equation gives the solution $1/u = r = 3m$.

Problem 2.17. (a) From the result of problem 2.4, $\frac{dr}{d\tau} = -(2m/r)^{1/2}$ for a particle falling from rest at infinity. Separating the variables and integrating:

$$\int_0^T d\tau = -\int_{2m}^0 \frac{r^{1/2}}{(2m)^{1/2}} dr$$

from which $T = 4m/3$.

(b) Use $g_{\mu\nu}u^\mu u^\nu = 1$. Inside the horizon this gives

$$1 = -\left(\frac{2m}{r} - 1\right)\dot{t}^2 + \left(\frac{2m}{r} - 1\right)^{-1}\dot{r}^2 - r^2\dot{\theta}^2 - r^2\sin^2\theta\dot{\phi}^2,$$

where the dot represents differentiation with respect to proper time. For this to be satisfied the positive term on the right-hand side must be ≥ 1. The equals sign will give the maximum proper time so

$$\left(\frac{2m}{r} - 1\right)^{-1}\dot{r}^2 = 1.$$

And hence

$$\frac{dr}{d\tau} = \pm\left(\frac{2m}{r} - 1\right)^{\frac{1}{2}}.$$

Taking the minus sign for inward motion and separating the variables gives

$$\int_0^T d\tau = -\int_{2m}^0 (2m/r - 1)^{-\frac{1}{2}} dr.$$

Making the substitution $r = 2m\cos^2\theta$ and integrating gives $T = \pi m$.

(c) Use equation (2.19) and set $L = 0$ and $E = 0$ as the conditions for a particle to be at rest at the horizon. This gives

$$\frac{dr}{d\tau} = -\left(\frac{2m}{r} - 1\right)^{\frac{1}{2}}.$$

which, as in part (b), integrates to $T = \pi m$. So an astronaut who wishes to survive for as long as possible inside the horizon should fire his rockets outside the horizon to reduce his speed as much as possible before crossing the horizon.

Problem 2.18. The time spent in pain is $\tau = \frac{2}{3}\left(\frac{\Delta r}{g_E}\right)^{\frac{1}{2}}$. To convert from geometric units to physical units use t(in metres) $= ct$(in seconds) and g(metres^{-1}) $= c^{-2}g$(ms^{-2}). On making these substitutions the c's cancel so the original expression is valid. Inserting $\Delta r = 2m$ and $g_E = 9.81$ ms^{-2} into the expression above gives $\tau = 0.3$s. The threshold of pain radius is $r_p = \left(\frac{2m\Delta r}{g_E}\right)^{\frac{1}{3}}$ in geometric units. Converting to physical units this becomes $r_p = \left(\frac{2GM\Delta r}{g_E}\right)^{\frac{1}{3}}$, after converting g and setting $m = GM/c^2$. For $M =$ one solar mass $r_p = 3.8 \times 10^3$ km. This is well beyond the horizon of a one solar mass black hole, which is about 3 km. The threshold of pain equals the Schwarzschild radius when $r_p = r_s$. This condition is realised when $M = \frac{c^3}{2G}\left(\frac{\Delta r}{g}\right)^{1/2} = 4.6 \times 10^4$ solar masses.

Problem 2.19. From the answer to problem 2.14 the time from crossing the horizon to hitting the singularity for a particle falling from a large distance is $4m/3$. Converting this expression to physical units and inverting to give mass in terms of time T gives

$$M = \frac{3c^3 T}{4G}.$$

For $T = 1$ yr we get $M = 5 \times 10^{12}$ solar masses. Note that only in the last 0.3 seconds will serious deformation occur, but 0.3 s is too short a time in which to suffer any pain.

Problem 2.20. In Rindler coordinates the metric is $d\tau^2 = \kappa^2\xi^2 dt^2 - d\xi^2$. Transforming to the coordinates (T, X) where

$$T = \xi \sinh \kappa t$$
$$X = \xi \cosh \kappa t$$
$$dT = d\xi \sinh \kappa t + dt\kappa\xi \cosh \kappa t$$
$$dX = d\xi \cosh \kappa t + dt\kappa\xi \sinh \kappa t.$$

Squaring and adding gives

$$dT^2 - dX^2 = \kappa^2\xi^2 dt^2 - d\xi^2.$$

So

$$d\tau^2 = dT^2 - dX^2.$$

Problem 2.21. On the curve $\xi = 1/a = $ constant

$$T = a^{-1} \sinh \kappa t$$

$$X = a^{-1} \cosh \kappa t.$$

Differentiating T and X with respect to τ gives

$$\frac{dT}{d\tau} = a^{-1} \kappa \cosh \kappa t \frac{dt}{d\tau},$$

and

$$\frac{dX}{d\tau} = a^{-1} \kappa \sinh \kappa t \frac{dt}{d\tau}.$$

Differentiating these expressions again with respect to τ gives the two non-zero components of the 4-acceleration:

$$\frac{d^2 T}{d\tau^2} = a^{-1} \kappa^2 \sinh \kappa t \left(\frac{dt}{d\tau}\right)^2 + a^{-1} \kappa \cosh \kappa t \frac{d^2 t}{d\tau^2}$$

$$\frac{d^2 X}{d\tau^2} = a^{-1} \kappa^2 \cosh \kappa t \left(\frac{dt}{d\tau}\right)^2 + a^{-1} \kappa \sinh \kappa t \frac{d^2 t}{d\tau^2}.$$

Now from the Rindler metric with $\xi = 1/a = $ constant

$$\left(\frac{dt}{d\tau}\right)^2 = \frac{1}{\kappa^2 \xi^2} = \frac{a^2}{\kappa^2} \quad \text{and} \quad \frac{d^2 t}{d\tau^2} = 0.$$

So the proper acceleration $A^\mu = (a \sinh \kappa\tau, a \cosh \kappa\tau, 0, 0)$ and $A^\mu A_\mu = -a^2$. So the magnitude of the proper acceleration is a.

Problem 2.22. We wish to find the limiting value of $\cos \psi$ as $r \to 0$, where

$$\cos \psi = \frac{[1 - (1 - 2m/r)b^2/r^2]^{1/2} - (2m/r)^{1/2}}{1 - (2m/r)^{1/2}[1 - (1 - 2m/r)b^2/r^2]^{1/2}}.$$

The velocity of a particle falling from rest at infinity, $(2m/r)^{1/2}$, has been inserted into equation (2.52). As $r \to 0$ the term $(1 - 2m/r)b^2/r^2$ dominates so

$$\cos \psi \to \frac{[(1 - 2m/r)b^2/r^2]^{1/2}}{(2m/r)^{1/2}[1 - (1 - 2m/r)b^2/r^2]^{1/2}} \to -\left(\frac{r}{2m}\right)^{1/2}.$$

Problem 2.23. From equation (2.55) we get

$$dU\,dV = \frac{1}{16m^2}\exp\left(\frac{v-u}{4m}\right)dv\,du.$$

But $v - u = 2r_*$ and r_* is given by (2.47), so we get

$$du\,dv = 16m^2 \exp\left[-\left(\frac{r}{2m} + \log\left(\frac{|r-2m|}{2m}\right)\right)\right]dU\,dV$$

$$= 16m^2 \left|\frac{2m}{r-2m}\right|\exp\left(-\frac{r}{2m}\right)dU\,dV$$

So, for $r > 2m$

$$\left(1 - \frac{2m}{r}\right)du\,dv = \frac{32m^3}{r}\exp\left(-\frac{r}{2m}\right)dU\,dV,$$

giving us the Kruskal metric (2.56), which is non-singular for all $r > 0$. Radial light rays satisfy $dU\,dV = 0$, hence $U = $ constant and $V = $ constant; the light cones in the $U - V$ plane are therefore parallel to the coordinate axes everywhere (see Fig. 2.6).

Problem 2.24. From the metric the proper time between a pair of events on the world line of a stationary clock is given by $d\tau = (1 - 2m/r)^{1/2}dt_{FF}$. So for a stationary clock at large r, $dt_{FF} = d\tau_\infty$. The wavelength of a light ray is proportional to its period, so $\lambda_r = (1 - 2m/r)^{1/2}\lambda_\infty$. Now, $z = (\lambda_\infty - \lambda_r)/\lambda_r$, so $1 + z = (1 - 2m/r)^{-1/2}$.

Problem 3.1. $\Delta = (r - r_+)(r - r_-)$ so $\Delta < 0$ for $r_- < r < r_+$.

Problem 3.2. From Section 3.2 the event horizon is a surface of constant r and constant t but its equatorial and polar circumferences are not the same. From Section 3.2.1 the equatorial circumference of an extreme Kerr black hole is $2\pi(r_+^2 + m^2)/r_+ = 4\pi m$ as $r_+ = m$. From the metric, the polar circumference of this black hole is $\int_0^{2\pi}\rho\,d\theta = \int_0^{2\pi}(r_+^2 + m^2\cos^2\theta)^{\frac{1}{2}}d\theta = m\int_0^{2\pi}(1 + \cos^2\theta)^{\frac{1}{2}}d\theta$. This integral cannot be evaluated in terms of elementary functions. However, we can estimate its value as follows. Expanding $(1+\cos^2\theta)^{\frac{1}{2}}$ by the binomial theorem gives $1+\frac{1}{2}\cos^2\theta - \frac{1}{8}\cos^4\theta + \frac{1}{16}\cos^6\theta + \cdots$ for the integrand. We can evaluate these four integrals: they give respectively $2\pi + \pi/2 - 3\pi/16 + 5\pi/48 = 2.42$. So the circumference is $\sim 2.42\pi m = 7.6m$. This is less than the equatorial circumference.

Problem 3.3. Putting $m = 0$ into the Boyer–Lindquist form of the Kerr metric gives

$$ds^2 = dt^2 - \frac{\rho^2}{r^2 + a^2} dr^2 - \rho^2 d\theta^2 - (r^2 + a^2) \sin^2 \theta d\phi^2.$$

We have to show that the coordinate transformation given in the question changes the flat space metric $dx^2 + dy^2 + dz^2$ into the spatial part of the metric given above. Differentiating the x coordinate transformation gives

$$dx = (r^2 + a^2)^{-\frac{1}{2}} r \sin \theta \cos \phi dr + (r^2 + a^2)^{\frac{1}{2}} \cos \theta \cos \phi d\theta$$
$$- (r^2 + a^2)^{-\frac{1}{2}} \sin \theta \sin \phi d\phi.$$

So

$$dx^2 = (r^2 + a^2)^{-1} r^2 \sin^2 \theta \cos^2 \phi dr^2 + (r^2 + a^2) \cos^2 \theta \cos^2 \phi d\theta^2$$
$$+ (r^2 + a^2) \sin^2 \theta \sin^2 \phi d\phi^2 + 2 \cos \theta \sin \theta \cos^2 \phi dr dr d\theta$$
$$- 2 \sin^2 \theta \cos \phi \sin \phi dr d\phi - 2(r^2 + a^2) \cos \theta \sin \theta \cos \phi \sin \phi d\theta d\phi.$$

Proceeding in the same way with the y and z transformations and adding them together gives finally

$$dx^2 + dy^2 + dz^2 = \frac{\rho^2}{(r^2 + a^2)} dr^2 + \rho^2 d\theta^2 + (r^2 + a^2) \sin^2 \theta d\phi^2.$$

Problem 3.4. We consider the identity $g_{03}^2 - g_{00} g_{03} = \Delta \sin^2 \theta$. Substituting for the metric coefficients and extracting the common factor gives

$$\frac{\sin^2 \theta}{\rho^4} [4m^2 a^2 r^2 \sin^2 \theta + (\Delta - a^2 \sin^2 \theta) A].$$

Inserting the expressions for A and Δ and using $\rho^2 = r^2 + a^2 \cos^2 \theta$ gives

$$\frac{\sin^2 \theta}{\rho^4} [4m^2 a^2 r^2 \sin^2 \theta + (\rho^2 - 2mr)(\rho^2(r^2 + a^2) + 2ma^2 r \sin^2 \theta)].$$

Finally the expression inside the square brackets reduces to $\rho^4 \Delta$ so the left-hand side becomes $\Delta \sin^2 \theta$ as required. The other identities follow in a similar manner.

Problem 3.5. For $r \gg a$

$$g_{00} = \frac{(\Delta - a^2 \sin^2 \theta)}{\rho^2} \rightarrow \frac{\Delta}{r^2} \rightarrow \left(1 - \frac{2m}{r}\right),$$

$$g_{03} = \frac{(2mar \sin^2 \theta)}{\rho^2} \rightarrow \frac{(2ma \sin^2 \theta)}{r}$$

$$g_{22} = -\rho^2 \rightarrow -r^2$$

$$g_{33} = -\frac{A \sin^2 \theta}{\rho^2} \rightarrow -r^2 \sin^2 \theta.$$

With these values of the metric coefficients the Kerr metric takes the form in equation (3.11).

Problem 3.6. The radius of the event horizon is $r_+ = m + (m^2 - a^2)^{1/2}$. If $a > m$ then r_+ becomes complex so there is no event horizon.

Problem 3.7. At the turning points of a bound orbit $d\theta/d\tau = 0$, so multiplying equation (3.16) through by $\sin^2 \theta_{TP}$ gives

$$Q \sin^2 \theta_{TP} = (1 - \sin^2 \theta_{TP})a^2(1 - E^2) \sin^2 \theta_{TP} + (1 - \sin^2 \theta_{TP})L_z^2.$$

Rearranging gives a quadratic equation in $\sin^2 \theta_{TP}$

$$a^2(1 - E^2) \sin^4 \theta_{TP} + [Q - a^2(1 - E^2) + L_z^2] \sin^2 \theta_{TP} - L_z^2 = 0.$$

The solution of which is

$$\sin^2 \theta_{TP}$$

$$= \frac{-(Q - a^2(1 - E^2) + L_z^2) \pm [(Q - a^2(1 - E^2) + L_z^2)^2 + 4a^2(1 - E^2)L_z^2]^{\frac{1}{2}}}{2a^2(1 - E^2)}.$$

For a bound orbit ($E < 1$) the term inside the square root is larger than the first term in the numerator, so to ensure that $\sin^2 \theta_{TP}$ is positive we take the plus sign. On taking the square root we see that there are two real values of $\sin \theta_{TP}$, that is two turning points for a bound orbit. Note that $Q = 0$ gives $\sin^2 \theta_{TP} = 1$: this is an orbit in the equatorial plane. The condition $L_z = 0$ gives $\sin^2 \theta_{TP} = 0$, which is an orbit with turning points at $\theta = 0$ and $\theta = \pi$, i.e. an orbit that passes over the two poles.

Problem 3.8. From equations (3.14) and (3.16)

$$\frac{d\phi}{dt} = \frac{d\phi/d\tau}{dt/d\tau} = \frac{L_z \Delta / \sin^2 \theta - aE\Delta + a[(r^2 + a^2)E - aL_z]}{AE - 2marL_z}.$$

At the horizon $\Delta = 0$, $r = r_+$ and $A = (2mr_+)^2$ so

$$\frac{d\phi}{dt} = \frac{a(2mr_+ E - aL_z)}{[(2mr_+)^2 E - 2mar_+ L_z]} = \frac{a}{2mr_+}$$

as required.

Problem 3.9. We have $\omega = \frac{u^3}{u^0} = \frac{g^{30}u_0 + g^{33}u_3}{g^{00}u_0 + g^{03}u_3} = \frac{g^{30}}{g^{00}}$ where the second equality follows because $L = -u_3 = 0$. Also from $u_3 = g_{03}u^0 + g_{33}u^3$ we get

$$\frac{u^3}{u^0} = \omega = -\frac{g_{03}}{g_{33}}.$$

Problem 3.10. The circumference of the event horizon of a Kerr black hole is $4\pi m$. (See Section 3.2.1.) Now velocity = circumference/period = $\frac{4\pi m}{2\pi/\omega}$. From equation (3.29) $\omega = 1/(2m)$ for an extreme black hole. Hence relative to an observer at infinity the velocity = 1 in geometric units, which is the speed of light.

Problem 3.11. From equation (3.29) the angular velocity of a Kerr black hole is given by $\omega_+ = \frac{a}{2mr_+}$. For $a \ll m$, $r_+ \sim 2m$ so $\omega_+ \sim a/(4m^2)$. By the usual definition, the moment of inertia I is given by $I = j/\omega_+$. So putting $j = ma$ and substituting for ω_+ gives $I = 4m^3$.

Problem 3.12. Setting $a = 0$ in equation (3.37): $(\frac{g_{33}}{-\Delta \sin^2 \theta})^{1/2} = (1 - \frac{2m}{r})^{-1/2}$ and $\omega = 0$. So the components of (u^μ_{ZAMO}) become $((1 - 2m/r)^{-1/2}, 0, 0, 0)$ which are the components of a hovering observer in the Schwarzschild metric.

Problem 3.13. The 4-velocity of a hovering observer is $(u^\mu) = (g_{00}^{-1/2}, 0, 0, 0)$ so the covariant components are

$$(u_\mu) = \left(g_{00}^{1/2}, 0, 0, g_{03}g_{00}^{-1/2}\right),$$

where $u_3 = g_{03}u^0$ has been used. Now evaluate the scalar product $u^\mu_{ZAMO} u_\mu$ in the Kerr frame and equate it to the energy per unit mass in a local frame.

Using equation (3.37) for u^{μ}_{ZAMO} with $\theta = \pi/2$ we get

$$\left(\frac{g_{33}}{-\Delta}\right)^{\frac{1}{2}} \left(g_{00}^{1/2} + \omega g_{03} g_{00}^{-1/2}\right) = \gamma.$$

Inserting $\omega = -g_{03}/g_{33}$ and using the identity $g_{03}^2 - g_{00}g_{33} = \Delta \sin^2 \theta$ we get

$$-\left(\frac{-\Delta}{g_{33}g_{00}}\right)^{\frac{1}{2}} = \gamma = \left(1 - \nu^2\right)^{-\frac{1}{2}}$$

or

$$\nu^2 = 1 + \frac{g_{33}g_{00}}{\Delta}.$$

Substituting for g_{33} and g_{00} and simplifying gives

$$\nu = \frac{2ma}{r\Delta^{1/2}}.$$

Problem 3.14. Substitute $a = m + \epsilon$ ($\epsilon > 0$) into $\Delta = r^2 + a^2 - 2mr$; then $\Delta = (r - m)^2 + 2m\epsilon + \epsilon^2$ which is positive for all r. Hence Δ does not go to zero and the energy of a ZAMO and the components of the 4-velocity of a ZAMO are always finite and real.

Problem 3.15. (i) Divide equation (3.49) by E_{ph}^2 and set $b = L_{ph}/E_{ph}$ to get

$$1 + \frac{2m}{r_{ph}^3}(b - a)^2 - \frac{1}{r_{ph}^2}(b^2 - a^2) = 0. \tag{4}$$

We have also from equation (3.50)

$$r_{ph} = 3m(b - a)/(b + a). \tag{5}$$

To obtain an equation for r_{ph} we eliminate b from these two simultaneous equations. Eliminating r_{ph} from (4) gives

$$(b + a)^3 = 27m^2(b - a). \tag{6}$$

(ii) We now use this result with (5) to find a simple quadratic for b:

$$r_{ph} = 3m\frac{(b - a)(b + a)^2}{(b + a)^3} = \frac{1}{9m}(b - a)^2$$

which re-arranges to

$$b^2 + 2ab + (a^2 - 9mr).$$

Thus, $b = 3(mr)^{1/2} - a$.

(iii) Substituting in (6) gives

$$\left(\frac{r_{ph}}{m}\right)^{3/2} - 3\left(\frac{r_{ph}}{m}\right) \pm \frac{2a}{m} = 0.$$

See Section 3.63 for the solution to this equation.

Problem 3.16. For purely radial motion of light the metric becomes

$$0 = g_{00}\left(\frac{dt}{d\lambda}\right)^2 + g_{11}\left(\frac{dr}{d\lambda}\right)^2.$$

But inside the static limit both the metric coefficients are negative so the condition cannot be satisfied.

Problem 3.17. Let u^μ be the 4-vector of the stationary particle. Then the 3-component of $u_\mu = g_{\mu\nu}u^\nu$ is

$$u_3 = g_{03}u^0 + g_{33}u^3 = g_{03}u^0$$

as $u^3 = d\phi/d\tau = 0$. Now, from the metric, the 4-velocity of a stationary particle is

$$u^\mu = \left(g_{00}^{-1/2}, 0, 0, 0\right).$$

So $u_3 = -L_z = g_{30}g_{00}^{-1/2}$. Finally

$$L_z = -\frac{2mar\sin^2\theta}{\rho(\Delta - a^2\sin^2\theta)^{1/2}}.$$

The denominator is zero at the static limit surface so $L_z \to \infty$ there. Therefore only a photon can remain static at this location. Inside the static limit surface the angular momentum becomes imaginary so a particle cannot be at rest.

Problem 3.18. Equation (3.59) written as a quadratic in a is

$$3a^2 \mp 8m^{\frac{1}{2}}r_{ms}^{\frac{1}{2}}a - (r_{ms}^2 - 6mr_{ms}) = 0$$

so

$$a = \frac{r_{ms}^{\frac{1}{2}}m^{\frac{1}{2}}}{3}\left(\pm 4 \mp \left(\frac{3r_{ms}}{m} - 2\right)^{\frac{1}{2}}\right).$$

To check the signs set $r_{ms} = m$. This gives $a = m$ (the prograde case) with first sign $+$ and second sign $-$. With $r_{ms} = 9m$ we get $a = m$ (the retrograde case) with the two signs reversed. So for a retrograde orbit with $r_{ms} = 7.5m$, $a = 0.48$ as required. For a prograde orbit with $r_{ms} = 4.12m$, $a = 0.53m$.

Problem 3.19. Let $1 - C = 3m/r > 0$ and $B = (a/m)(m/r)^{3/2}$. Then

$$\Delta\tau_+^2 - \Delta\tau_-^2 = \frac{(C + 2B)(1 - B)^2 - (C - 2B)(1 + B)^2}{(1 - B^2)^2}.$$

The numerator reduces to $4B(1 - C + B^2) > 0$ and the denominator is positive. Thus $\Delta\tau_+^2 > \Delta\tau_-^2$ and since both are positive $\Delta\tau_+ > -\Delta\tau_-$. So the twin in a co-rotating orbit does indeed age more quickly, relative to a distant observer, than the twin orbiting at the same radius in the opposite sense.

Problem 3.20. We are given $r = 4.12m$ for the last stable orbit of a black hole with $a = 0.53m$ and $T = 2\pi\left(\frac{r^{3/2}}{m^{1/2}} \pm a\right)$. Substituting for r and a gives $T = 2\pi m[(4.12)^{3/2} + 0.53]$. Putting $m = GM/c^2$ and $t = cT'$ to convert to physical units and using $M = 4 \times 10^6 M_\odot$ gives $T = 18.4$ minutes.

Problem 3.21. We use the expression for a as a function of r_{ms} derived in problem 3.18 to obtain values of r_{ms} for a range of values of a. We then use these values of r in equation (3.58).

Problem 3.22. Equation (3.54), after substitution for E^2 from equation (3.58), becomes

$$L^2 = \frac{2m}{3r_{ms}}\left(3r_{ms}^2 - a^2\right). \tag{7}$$

Now solving equation (3.59) as a quadratic in a (see problem 3.18) gives

$$a = \frac{m^{\frac{1}{2}}r_{ms}^{\frac{1}{2}}}{3}\left[\pm 4 \mp \left(3\frac{r_{ms}}{m} - 2\right)^{\frac{1}{2}}\right].$$

Squaring a and substituting into (7) gives

$$L^2 = \left(\frac{4m^2}{27}\right)\left[12\left(\frac{r_{ms}}{m}\right) - 7 + 4\left(3\frac{r_{ms}}{m} - 2\right)^{\frac{1}{2}}\right].$$

Finally taking the square root gives

$$L = \pm\frac{2m}{3\sqrt{3}}\left[1 + 2\left(3\frac{r_{ms}}{m} - 2\right)^{\frac{1}{2}}\right].$$

Problem 3.23. (a) For the co-rotating case in the extreme Kerr limit $r_{ms} = m$ so $L = 2m/\sqrt{3}$.
(b) For the counter-rotating case $r_{ms} = 9m$ so $L = -22m/(3\sqrt{3})$.

Problem 3.24. Subsitute $r = 2m$ into equation (3.63). This gives $a = m/\sqrt{2} = 0.7071m$.

Problem 3.25. Let $x = r/m - 1$. Then

$$D = m \int_{x_+}^{X} \frac{x+1}{(x^2 - 2\delta)^{1/2}} dx$$

$$= \left[m \left(x^2 - 2\delta \right)^{1/2} \right]_{x_+}^{X} + m \left[\log \left\{ \left(x^2 - 2\delta \right)^{1/2} + x \right\} \right]_{x_+}^{X}.$$

With $X = x_{ph} = (8\delta/3)^{1/2}$ this becomes

$$D = \left(\frac{2\delta}{3} \right)^{1/2} + m \log \left[\frac{(2\delta/3)^{1/2} + (8\delta/3)^{1/2}}{(2\delta)^{1/2}} \right].$$

Letting $\delta \to 0$ gives

$$D = (m/2) \log 3.$$

Similarly with $X = x_{mb}$ we get

$$D = m (2\delta)^{1/2} + m \log \left[\frac{(2\delta)^{1/2} + 2(\delta)^{1/2}}{(2\delta)^{1/2}} \right]$$

which in the limit $\delta \to 0$ gives

$$D = m \log(1 + \sqrt{2}).$$

Problem 3.26. For large r equation (3.67) becomes

$$E^2 = \frac{r \left(r^2 - 2mr \right)^2}{r^2 \left(r^3 - 3mr^2 \right)} = \frac{\left(1 - \frac{2m}{r} \right)^2}{\left(1 - \frac{3m}{r} \right)}$$

which is the Schwarzschild result for the energy of a particle in a circular polar orbit.

Problem 3.27. (a) In equation (3.28) the dominant term in the denominator for large r is r^4 so the equation reduces to $\omega_p = 2ma/r^3$.

(b) In geometrical units $\omega_p = 2ma/r^3$. Now $c \times t(\text{seconds}) = t(\text{metres})$ so $\omega(\text{seconds}^{-1}) = c \times \omega(\text{metres}^{-1})$. We also have $m = GM/c^2$ and $a = J/Mc$. So

$$\omega(\text{s}^{-1}) = \frac{2GJ}{c^2 r^3}.$$

(c) We have

$$\omega_p = \frac{2ma}{r^3} \times \frac{m^3}{m^3} = \frac{2a}{m^2 \left(r/m\right)^3},$$

so for a given r/m the precessional frequency is inversely proportional to the mass of the black hole squared.

Problem 3.28. For a polar orbit about the Earth we can use the expression for the precession angular velocity given in SI units in problem 3.27

$$\frac{d\phi}{dt} = \frac{2GJ}{c^2 r^3},$$

with $J = 0.33 M_\oplus R_\oplus^2 \Omega_\oplus = 0.33 \times 6 \times 10^{24} \times (6.37 \times 10^6)^2 \times 7.3 \times 10^{-5} = 5.86 \times 10^{33}\,\text{Js}$. So in one year $\Delta\phi = (2 \times 6.7 \times 10^{-11} \times 5.86 \times 10^{33} \times 3.16 \times 10^7)/(9 \times 10^{16} \times (12 \times 10^6)^3) = 1.6 \times 10^{-7}$ radians or 33 milli-arc seconds.

Problem 3.29. For a polar orbit $L_z = 0$ so equation (3.16) gives for the polar orbit

$$\rho^2 \frac{d\phi}{d\tau} = \frac{2marE}{\Delta}.$$

For a zero angular momentum particle in the equatorial plane we have $L_z = 0$ and $\theta = \pi/2$ so for this case equation (3.16) gives

$$r^2 \frac{d\phi}{d\tau} = \frac{2marE}{\Delta}.$$

When $r \gg a$ we have $\rho \to r$ and these two expressions tend to the same value.

Problem 3.30. The condition $V_r = 0$ gives

$$Q = \left[E^2(r^2 + a^2)^2 - \Delta a^2 E^2 - r^2 \Delta\right]/\Delta.$$

Substituting for E^2 from equation (3.67) gives, after simplification, the required expression. The final part of the question follows immediately on setting $a = 0$. In this limiting case $Q = L^2$.

Problem 3.31. From equation (3.38) setting $L_z = 0$

$$\rho^4 \left(\frac{dr}{d\tau} \right)^2 = E_{ph}^2 (r^2 + a^2)^2 - \Delta(a^2 E_{ph}^2 + Q) = R(r).$$

The conditions for an orbit at constant r are $R(r) = 0$ and $dR/dr = 0$. The latter condition gives

$$2r(r^2 + a^2) - (r - m)(a^2 + Q/E^2) = 0.$$

So

$$\frac{Q}{E_{ph}^2} = \frac{(r - m)a^2 - 2r(r^2 + a^2)}{r - m}$$

and for $a = m$ and $r_{ph} = (1 + \sqrt{2})m$ we get $Q/E_{ph}^2 = 11 + 8\sqrt{2}$ as required.

Problem 3.32. The Killing vector $(k^\mu) = (1, 0, 0, 0)$ has covariant components $k_\mu = g_{\mu\nu} k^\nu$ so $k_0 = g_{00} k^0 + g_{03} k^3 = g_{00} k^0$ as $k^3 = 0$. So the scalar product $k^\mu k_\mu = g_{00}$. At the static limit surface g_{00} changes from being positive outside to negative inside. Therefore the vector k^μ is spacelike inside the static limit surface.

Problem 3.33. Setting the index $\mu = 0$ in $p_\mu = g_{\mu\nu} p^\nu$ for a photon gives

$$p_0 = E_{ph} = g_{00} p^0 + g_{03} p^3 = g_{00} \frac{dt}{d\lambda} + g_{03} \frac{d\phi}{d\lambda}$$

or

$$E_{ph} = \frac{dt}{d\lambda}(g_{00} + g_{03}\Omega), \tag{8}$$

where $\Omega = \frac{d\phi}{dt}$. Now from equation (3.48) $\Omega = w \pm \frac{r^2 \Delta^{1/2}}{A}$ for a tangentially projected photon and $w = -\frac{g_{03}}{g_{33}}$. Substituting these expressions into (8) and using the identity $g_{03}^2 - g_{00} g_{33} = \Delta \sin^2 \theta$ gives for motion in the equatorial plane

$$E_{ph} = \frac{dt}{d\lambda} \left(\frac{\Delta r^2 \pm 2mar\Delta^{1/2}}{A} \right). \tag{9}$$

The timelike component of $p^\mu = g^{\mu\nu} p_\nu$ is $\frac{dt}{d\lambda}$:

$$\frac{dt}{d\lambda} = \frac{A E_{ph} - 2mar L_{ph,z}}{r^2 \Delta}.$$

Substituting into (9) and collecting the terms containing E_{ph} on the left-hand side gives, after some cancellations,

$$E_{ph} = \frac{2marL_{ph} + |L_{ph}|r^2\Delta^{1/2}}{A}, \tag{10}$$

where L_{ph} is positive for co-rotation and negative for contra-rotation. A particle at the horizon ($\Delta = 0$) has $E_{ph} = \frac{2mar_+L_{ph,z}}{(2mar_+)^2} = \frac{aL_{ph,z}}{2mr_+} = \omega_+L_{ph,z}$. Inserting $\dot{r} = 0$ into equation (3.76) and setting the $r^2\Delta$ term inside the square root to zero, as we are treating a photon, reduces (3.76), after some simplification, to equation (10). If L is negative then equation (10) tells us that inside the ergosphere, where $r^2\Delta^{1/2} < 2mar$, E_{ph} is negative.

Problem 3.34. The innermost stable orbit for a co-rotating particle around an extreme Kerr black hole is at $r_{ms} = m$. The angular momentum gained through accretion of a mass dm from the last stable orbit of an extreme black hole, from equation (3.84) with $x_{ms} = 1$, is given by $dj = 2mdm$. We get j as a function of m by integrating:

$$\int_J^{J'} dj = \int_M^{M'} 2mdm,$$

where the lower limits, J and M are the angular momentum and mass of an extreme Kerr black hole, which was our starting point. Evaluating the integral gives $J' - M'^2 = J - M^2$. But for an extreme Kerr black hole $J = M^2$, so $J' = M'^2$ and the extreme Kerr limit has not been exceeded.

Problem 3.35. From equation (3.88) $r_{ms}m = $ constant, so $9m_K^2 = 6m_S^2$ hence $m_K = \sqrt{2/3}m_s$. So the fractional increase in mass of the black hole is $\sqrt{3/2} - 1 = 22.5\%$.

To find the quantity of rest mass that has been accreted we employ the method used above. Equation (3.83) with $r_{ms} = 9m_K^2/m$ gives

$$dm_0 = \frac{dm}{\left(1 - \frac{2m^2}{27m_K^2}\right)^{\frac{1}{2}}}.$$

Now integrating

$$\Delta m_0 = a \int_{m_K}^{\sqrt{3/2}m_K} \frac{dm}{(a^2 - m^2)^{1/2}}$$

where here $a = \sqrt{27/2} m_K$. Evaluating the integral by using the substitution $m = a \sin \theta$ gives

$$\Delta m_0 = \left(\frac{27}{2}\right)^{\frac{1}{2}} \left[\sin^{-1}\left(\frac{1}{3}\right) - \sin^{-1}\left(\frac{2}{27}\right)^{\frac{1}{2}}\right] m_K$$

so $\Delta m_0 = 0.236 m_K$. This is larger than the increase in the mass of the black hole by $0.011 m_K$. So 4.7% of the rest mass accreted has been radiated away. Thus, a black hole accreting from a contra-rotating disc will not be as luminous as one accreting from a co-rotating disc.

Problem 3.36. Initially $a = 0$ so the rate at which energy is dissipated in the accretion disc by a mass Δm, on its way to the last stable orbit, is $0.057 \Delta m/\Delta t$. (See Section 2.3.12.) Finally, with $a = 0.998m$ say, the rate of energy dissipation in the disc rises to $0.3 \Delta m/\Delta t$. The luminosity is proportional to the rate of energy dissipation so, assuming the mass flow rate is constant, the luminosity increases by a factor of 5.3.

Problem 3.37. To answer this question we have to show that the integral of ds is positive around the circle $r = a\delta$, $T = $ constant in the $Z = 0$ plane. Inserting these conditions into the metric in Kerr–Schild coordinates gives

$$ds^2 = -dX^2 - dY^2 - \frac{2ma^3\delta^3}{(a\delta)^4}$$
$$\times \left[\frac{-a\delta(XdX + YdY) - a(XdY - YdX)}{(a\delta)^2 + a^2}\right]^2. \tag{11}$$

From the coordinate transformations in Section 3.18 we get, with $\theta = \pi/2$:

$$dX = (-r \sin \tilde{\phi} + a \cos \tilde{\phi})d\tilde{\phi} = -Y d\tilde{\phi} \tag{12}$$
$$dY = (r \cos \tilde{\phi} + a \sin \tilde{\phi})d\tilde{\phi} = X d\tilde{\phi}. \tag{13}$$

Substituting these results into (11) gives

$$ds^2 = \left[-r^2 - a^2 - \frac{2m}{\delta}\left[\frac{-X^2 - Y^2}{(a\delta)^2 + a^2}\right]^2\right] d\tilde{\phi}^2.$$

With delta negative and small ds^2 is positive and so the interval is timelike.

Problem 3.38. The event horizon is given by $\Delta = r^2 + a^2 + q^2 - 2mr = 0$. Solving this quadratic in r for the radius of the horizon gives $r_+ = m \pm (m^2 - a^2 - q^2)^{1/2}$. So r_+ is real only when $m^2 > a^2 + q^2$.

Problem 4.1. (i) We evaluate the scalar product $\chi^\mu \chi_\mu$, where $(\chi^\mu) = (1, 0, 0, \omega)$. Now $\chi_0 = g_{00}\chi^0 + g_{03}\chi^3$ and $\chi_3 = g_{30}\chi^0 + g_{33}\chi^3$ so $\chi_\mu = (g_{00} + g_{03}\omega, 0, 0, g_{30} + g_{33}\omega)$ and $\chi^\mu \chi_\mu = g_{00} + g_{03}\omega + g_{03}\omega + g_{33}\omega^2$. Now $\omega = -g_{03}/g_{33}$ so $\chi^\mu \chi_\mu = -(\Delta \sin^2 \theta)/g_{33}$. But at the horizon $\Delta = 0$, so χ^μ is null at the horizon.

(ii) The metric in ingoing Eddington–Finkelstein coordinates is

$$ds^2 = \left(1 - \frac{2m}{r}\right) dv^2 + 2dvdr - r^2 d\tilde{\omega}^2.$$

$\chi^\mu = (1, 0, 0, 0)$ is null on the horizon so $\chi^\nu(\nabla_\nu \chi^\mu) = \kappa\chi^\mu$ reduces to

$$\nabla_0 \chi^0 = \kappa.$$

Now $\nabla_\nu \chi^\mu = \chi^\mu_{,\nu} + \Gamma^\mu_{\nu\rho}\chi^\rho$ so $\nabla_0 \chi^0 = \Gamma^0_{00}\chi^0 = \kappa$. From the definition of $\Gamma^\mu_{\nu\rho}$ (see equation (1.10)) with $g_{00} = (1 - 2m/r)$ and $g_{01} = 1$, $\Gamma^0_{00} = \frac{1}{2}g^{01}\frac{\partial g_{00}}{\partial r} = \frac{m}{r^2}$. At the horizon, $r = 2m$ so finally $\kappa = 1/(4m)$.

Problem 4.2. To verify that $m = \kappa A_h/4\pi + 2\omega_+ j$. Substituting for κ, A_h and j:

$$\kappa A_h/4\pi + 2\omega_+ j = \frac{(m^2 - a^2)^{1/2}}{2mr_+}2m[m + (m^2 - a^2)^{1/2}] + \frac{2ma^2}{2mr_+}$$

which simplifies to

$$\frac{m(m^2 - a^2)^{1/2}}{r_+} + \frac{m^2}{r_+} = \frac{m[m + (m^2 - a^2)^{1/2}]}{r_+} = m.$$

This verifies Smarr's formula.

Problem 4.3. The solution to this problem follows closely the treatment for the increase in surface area due to the capture of a massive particle given in Section 4.2.1. Equation (4.7) gives

$$\delta A_h = \frac{16mr_+\pi}{(m^2 - a^2)^{1/2}} (\delta m - \omega_+ \delta j).$$

For a photon $\delta m = E_{ph}$ and from Section 3.10.3 the impact parameter for a photon in a circular orbit around an extreme Kerr black hole is $b = L_{ph}/E_{ph}$ but $b = 2m$ for such a photon so $L_{ph}/E_{ph} = 2m$ and the change in angular momentum $\delta j = 2mE_{ph}$. Using these results the change in area becomes

$$\delta A_h = \frac{16m\pi E_{ph}}{(m^2 - a^2)^{1/2}}(r_+ - a) = \frac{16m\pi E_{ph}}{(m^2 - a^2)^{1/2}}\left[(m^2 - a^2)^{1/2} + \varepsilon\right]$$

as $r_+ - a = (m^2 - a^2)^{1/2} + \varepsilon$. (See Section 4.2.1.) So finally the change in area as $a \to m$ is

$$\delta A_h = 16m\pi E_{ph}.$$

So the area increases.

Problem 4.4. The total surface area of the two Schwarzschild black holes is $A_S = 2 \times 16\pi m^2$. The surface area of an extreme Kerr black hole is $A_K = 8\pi M^2$. So the change in area is $A_K - A_S = 8\pi(M^2 - 4m^2)$. Now the area theorem requires that $\Delta A \geq 0$ or $M - 2m \geq 0$. But M cannot be greater than $2m$ by conservation of energy, so we must have $M = 2m$. There is therefore no energy to radiate.

Problem 4.5. Starting from equation (4.5) $A_h = 8\pi m[m + (m^2 - a^2)^{1/2}]$, rearrange and square to get $(\frac{A_h}{8\pi m} - m)^2 = m^2 - a^2$. Evaluating the left-hand side and putting $a = j/m$ gives $(\frac{A_h}{8\pi m})^2 + \frac{j^2}{m^2} = \frac{2A_h}{8\pi}$. Finally multiply though by $4\pi m^2/A_h$ to get $\frac{A_h}{16\pi} + \frac{4\pi j^2}{A_h} = m^2$.

Problem 4.6. From Section 4.2.4 $m_{rot} = j^2(A_h/4\pi)^{-3/2}$, $j = I\omega_+$ and for $a \ll m$, $A_h \sim 16\pi m^2$ so $m_{rot}, = j^2(A_h/4\pi)^{-3/2} = I^2\omega^2/8m^3 = (I/8m^3)I\omega_+^2 = \frac{1}{2}I\omega_+^2$.

Problem 4.7. The total area of the event horizons of the two Schwarzschild black holes before the merger is

$$A_{initial} = 2 \times 4\pi(2m_i)^2 = 32\pi m_i^2.$$

The area of the resultant black hole is

$$A_{final} = 4\pi(2m_f)^2 = 16\pi m_f^2.$$

To maximise the energy dissipated in the merger the area of the final black hole must be as low as the Hawking area theorem allows, which is $A_{final} = A_{initial}$. So equating areas gives

$$M_f = \sqrt{2}m_i.$$

Thus the fraction of the initial mass lost in the merger is $1 - 2^{-1/2} = 0.29$.

Problem 4.8. The total horizon area before the merger is

$$A_i = 2 \times 4\pi(m_K)^2$$

and the area of the final Schwarzschild black hole is

$$A_f = 4\pi(2m_s)^2.$$

Equating the area before to the area after gives $m_K = m_S$. So half of the initial mass has been radiated away.

Problem 4.9. In the limit as $r \to \infty$ the first term inside the curly brackets reduces to ω^2 and the other two terms tend to zero. So the Teukolsky equation simplifies to

$$\frac{d^2u}{dr_*^2} = -\omega^2 u,$$

which has equation (4.24) as its solution. In the limit as $r \to r_+$, $\Delta \to 0$ so in this case the Teukolsky equation reduces to

$$\frac{d^2u}{dr_*^2} = \frac{\left[(r_+^2 + a^2)\omega - aM\right]^2 u}{\left(r_+^2 + a^2\right)^2} = -(\omega - \omega_+ M)^2 u$$

as $(r_+^2 + a^2) = 2mr_+$ and $\omega_+ = a/2mr_+$. Thus in this limit the Teukolsky equation has equation (4.25) as its solution. Putting $a = 0$ in the Teukolsky equation gives

$$\frac{d^2u}{dr_*^2} + \omega^2 u - \frac{l(l+1)(1-2m/r)u}{r^2} - \frac{(1-2m/r)^2}{r^2} - \frac{d}{dr_*}\left(\frac{(1-2m/r)}{r}\right)u = 0.$$

As

$$\frac{d}{dr_*} = \frac{\Delta}{r^2 + a^2}\frac{d}{dr}$$

the last term becomes $(1 - 2m/r)^2/r^2 - 2m(1 - 2m/r)/r^3$; so finally the Teukolsky equation reduces to

$$\frac{d^2u}{dr_*^2} + \omega^2 u - \frac{(1-2m/r)[l(l+1)+2m/r]}{r^2} = 0$$

which is the Regge–Wheeler equation.

Problem 4.10. We are given that $W = u'v - v'u = \text{constant}$ for any two independent solutions u and v of a second order differential equation. A solution of equation (4.17) in the limit as $r_* \to -\infty$ is $u \sim e^{-i\omega r_*}$ (equation 4.18). Another solution is $v = u^*$. So differentiating u and v with respect to r_* gives $u' = -i\omega e^{-i\omega r_*}$ and $v' = i\omega e^{i\omega r_*}$. Hence $W = -2i\omega$. In the limit as $r_* \to +\infty$ the solutions to (4.17) takes the form $u \sim a_{out}e^{i\omega r_*} + a_{in}e^{-i\omega r_*}$ and $v = u* \sim a_{out}e^{-i\omega r_*} + a_{in}e^{i\omega r_*}$. Differentiating with respect to r_* to get u' and v' we get a second expression for W:

$$W = 2i\omega(a_{out}^2 - a_{in}^2).$$

Equating these two expressions gives $1 + a_{out}^2 = a_{in}^2$ the relation quoted in the text as equation (4.20).

In a similar way we can repeat this calculation for the limiting solutions of the Teukolsky equation to get the relation (4.26).

Problem 4.11. For a black hole radiating as a blackbody at temperature T we can write $kT \sim h\nu = hc/\lambda$ where λ is the wavelength at the peak of the blackbody spectrum. Now λ will be of order of the dimensions of the black hole (the only length scale in the problem), hence of order the Schwarzschild radius. So substituting $\lambda = 2GM/c^2$ into the above expression gives $T \sim hc^3/2kGM$.

Problem 4.12. To show that $U_{\omega'}^{(+)} = \frac{1}{2(\pi|\omega'|)^{1/2}}e^{-i|\omega'|\eta}\xi^{i\omega'/a}$ is a solution of the equation

$$-\frac{1}{a^2\xi^2}\frac{\partial^2\Phi}{\partial\eta^2} + \frac{1}{\xi}\frac{\partial\Phi}{\partial\xi} + \frac{\partial^2\Phi}{\partial\xi^2} = 0.$$

The factor $2(\pi|\omega'|)^{-1/2}$ will cancel out so we can ignore it. Differentiating $U_\omega^+ = e^{-i|\omega'|\eta}\xi^{i\omega'/a}$ twice with respect to η gives

$$\frac{\partial^2 U}{\partial\eta^2} = i^2\omega'^2 e^{i\omega'\eta}\xi^{i\omega'/a}.$$

Differentiating $U_\omega^{(+)}$ with respect to ξ gives

$$\frac{\partial U}{\partial\xi} = e^{-i\omega'\eta}\left(i\omega'/a\right)\xi^{-1+i\omega'/a},$$

and differentiating a second time gives

$$\frac{\partial^2 U}{\partial\xi^2} = e^{-i\omega'\eta}\left(i\omega'/a\right)\left(i\omega'/a - 1\right)\xi^{-2+i\omega'/a}.$$

Now substituting these three partial derivatives into the wave equation the quantity $e^{-i|\omega'|\eta}\xi^{-2+i\omega'/a}$ cancels leaving $-\frac{i^2\omega'^2}{a^2} + \frac{i\omega'}{a} + \frac{i^2\omega'^2}{a^2} - \frac{i\omega'}{a} = 0$ so equation (4.34) is a solution of equation (4.33).

Problem 4.13. From the definition of r_* given in equation (4.16)

$$\frac{r_*}{2m} - 1 = \frac{r}{2m} + \log\left(\frac{r}{2m} - 1\right) - 1.$$

So near $r = 2m$

$$\frac{r_*}{2m} - 1 \sim \log\left(\frac{r}{2m} - 1\right)$$

and

$$\frac{r}{2m} - 1 \sim 1 - \frac{2m}{r} \sim \kappa^2 \xi^2$$

where the second equality comes from equation (2.41). Combining (4.13) and (4.13) gives

$$\kappa^2 \xi^2 \sim \exp\left(\frac{r_*}{2m} - 1\right).$$

Now,

$$U = -\exp\left(\frac{-u}{2m}\right)$$

and $u = t - r_*$. So

$$U = -e^{r_*/4m} e^{-\kappa t}.$$

Using (4.13) and $\kappa = 1/4m$ we get $U \propto \xi e^{-\kappa t}$.

Similarly the Kruskal coordinate $V = \exp(v/4m)$ near to $r = 2m$ becomes $V \propto \xi e^{\kappa t}$. The transformation from Rindler coordinates (t, ξ) to Minkowski null coordinates $u = T - X$ and $v = T + X$ is given by $u = \xi e^{-\kappa \tau}$ and $v = \xi e^{\kappa \tau}$ using (2.43). Thus near $r = 2m$ the metric in Kruskal coordinates has approximately the Minkowski form in null coordinates.

Problem 4.14. The entropy of a black hole is given by $S = \frac{k_B c^3 A_h}{4\hbar G}$, where A_h is the area of the horizon. For a spherical black hole the horizon area is $4\pi(2m_{pl})^2$ in geometrical units. In physical units this becomes $4\pi(2GM_{pl}/c^2)^2$ and $M_{pl} = (\hbar c/G)^{1/2}$. So $A_h = 16\pi\hbar G/c^3$ and hence $S = 4\pi k_B$.

Problem 4.15. Substituting $M = 2 \times 10^{30}$ kg into the equation given in the text gives $t_h = [8.33 \times 10^{19} \times (2 \times 10^{30}/10^{12})^3]/3.16 \times 10^7 = 2.2 \times 10^{67}$ years.

Problem 4.16. A black hole with a mass of 1.6×10^{12} kg has a lifetime of $t_h \sim 2.5 \times 10^9 \times (1.6)^3$ y or 1.02×10^{10} y. Assume a typical gamma ray energy is 1 MeV. This corresponds to a temperature of $\sim 10^{10}$ K. Using

$T_h \sim 10^{23}/M$ K gives $M \sim 10^{13}$ K so such a black hole will have an age of order the age of the universe.

Problem 4.17. We have, $kT = \frac{\hbar c^3}{8\pi GM}$. Setting $M = (G/\hbar c)^{-\frac{1}{2}}$, the Planck mass, gives $kT = \frac{1}{8\pi}(\frac{\hbar c^5}{G})^{\frac{1}{2}}$ so $T = 7.78 \times 10^7$ K. We also have $S = \frac{kc^3}{4\hbar G}A_h$ and $A_h = 1\text{m}^2$ so $S = k(\frac{c^3}{4\hbar G}) = 1.3 \times 10^{46}$ JK^{-1}.

Problem 4.18. The relationship between M and T is $M = \frac{\hbar c^3}{8\pi GkT}$ (see Section 4.6.1) so

$$\frac{d(Mc^2)}{dT} = -\frac{\hbar c^5}{8\pi GkT^2}.$$

Expressing T in terms of M the result follows.

Problem 4.19. To obtain the surface gravity of a Reissner–Nordström black hole we find the acceleration of a hovering observer by repeating the steps in Section 2.8 for the Schwarzschild case, only this time with $g_{00} = g_{11}^{-1} = (1 - 2m/r + q^2/r^2)$. This gives for the acceleration $a = (m/r^2 - q^2/r^3)(1 - 2m/r + q/r^2)^{-\frac{1}{2}}$. To get the acceleration measured by an observer at infinity we use equation (2.38) from Section 2.8.2 with the Reissner–Nordström redshift factor replacing the Schwarzschild factor. This gives for the surface gravity $\kappa = \frac{(mr_+ - q^2)}{r_+^3}$, where $r_+ = m + (m^2 - q^2)^{\frac{1}{2}}$. An extreme Reissner–Nordström black hole has $q = m$ and $r_+ = m$ hence $\kappa = 0$. And so $T = \hbar\kappa/(2\pi ck) = 0$.

Problem 4.20. Let $x = M_{eq}/M_i$ and find the value of x that maximises $(1-x)x^4$ by differentiating with respect to x and setting to zero. This gives

$$-x^4 + 4(1-x)x^3 = 0.$$

Thus $x = 4/5$. Substituting this value of x back in to the original expression gives $(1 - 4/5)(4/5)^4 = 0.082$.

Problem 5.1. The metric is

$$d\tau^2 = dt^2 - dl^2 - (b_0^2 + l^2)(d\theta^2 + sin^2\theta d\phi^2).$$

In the $\theta = \pi/2$ plane at constant t the spatial metric is

$$dL^2 = dl^2 + (b_0^2 + l^2)d\phi^2. \tag{14}$$

The metric of Euclidean space is

$$dL^2 = dz^2 + dr^2 + r^2 d\phi^2.$$

On the surface $z = z(r)$ this becomes

$$dL^2 = \left[1 + \left(\frac{dz}{dr}\right)^2\right] dr^2 + r^2 d\phi^2. \tag{15}$$

Equations (14) and (15) are the same if

$$(1 + z'^2) dr^2 = dl^2 \quad \text{and} \quad r^2 = b_0^2 + l^2.$$

Thus

$$1 + z'^2 = \left(\frac{dl}{dr}\right)^2 = \left(\frac{r}{l}\right)^2.$$

Integrating this equation from $r = b_0$ at $z = 0$ gives

$$z = b_0 \cosh^{-1}\left(\frac{r}{b_0}\right).$$

The circumference of the throat located at $z = 0$, $r = b_0$ is $2\pi b_0$.

Problem 5.2. Let the falling body have 4-velocity components u^μ. Since the metric is static we have, as usual, $u_0 = E$. Now, a hovering (stationary) observer has 4-velocity $u_H^0 = dt/d\tau = e^{-\Phi}$ so

$$u_\mu u_H^\mu = E e^{-\Phi} = \gamma = (1 - v^2)^{-1/2}$$

from which

$$v = \left(1 - \frac{e^{2\Phi}}{E^2}\right)^{\frac{1}{2}}.$$

The speed v is that measured by the local stationary observer, hence $v = $ (proper distance)/(proper time of the local observer) $= dl/e^\Phi dt$. The time to traverse the wormhole is obtained by integration.

Problem 5.3. Starting from the metric

$$d\tau^2 = dt^2 - \frac{dr^2}{1 - b_0^2/r^2} - (b_0^2 + l^2)(d\theta^2 + \sin^2\theta d\phi^2)$$

we set $\theta = \pi/2$ and $d\tau^2 = 0$. The constants of the motion are $u_0 = E_{ph}$ and $u_3 = -L_{ph}$. Then $g_{\mu\nu} u^\mu u^\nu = 0$ gives

$$\left(\frac{dr}{d\lambda}\right)^2 = E_{ph}^2 \left[1 - \frac{L_{ph}^2}{E_{ph}^2 r^2}\right]\left(1 - \frac{b_0^2}{r^2}\right).$$

For $L_{ph}/E_{ph} > b_0$ the first bracket on the right is zero for $r > b_0$. This is the point of closest approach: the radial coordinate of the light ray increases past this turning point and the light ray escapes to infinity. For $L_{ph}/Eph <$ b_0 the light ray enters the wormhole. The dividing value gives the impact parameter $L/E = b_0$. The capture cross-section is then πb_0^2 because r is an area coordinate.

Problem 5.4. The 4-velocity of a static observer is $u^\mu = (e^{-\Phi}, 0, 0, 0)$ and the 4-acceleration is

$$a^\mu = \frac{Du^\mu}{d\tau} = \frac{du^\mu}{d\tau} + \Gamma^\mu_{\rho\sigma} u^\rho u^\sigma.$$

We have $a^\mu u_\mu = a^0 u_0 = 0$ so $a^0 = 0$. Also, $a^2 = a^3 = 0$ by symmetry. Thus

$$a^1 = \Gamma^1_{00}(u^0)^2$$

$$= \frac{1}{2} g^{11} \left(-\frac{\partial e^{2\Phi}}{\partial r} \right) (u^0)^2$$

$$= -\left(1 - \frac{r}{r} \right) \Phi'.$$

The acceleration a is now obtained from $a^2 = a^\mu a_\mu$ as

$$a = \left(1 - \frac{b}{r} \right)^{\frac{1}{2}} \Phi'.$$

Problem 5.5. We have $u^\mu v_\mu = \gamma$ and $g^{\mu\nu} v_\mu v_\nu = 1$ from which the result follows with some simple algebra.

Problem 5.6. If $\rho > 0$ and $p = -\rho$ then $p + \rho = 0$ which is consistent with the weak energy condition. In the frame of an observer moving at 4-velocity v^μ relative to the fluid the energy density is

$$T^{0'0'} = \gamma^2(\rho + v^2 p) = \gamma^2(1 - v^2)\rho = \rho = T^{00}. \tag{16}$$

Problem 6.1. To simplify slightly we put $l = 1$. The factors of l can be reinstated at the end on dimensional grounds. We begin by computing the differentials from (6.7):

$$dT_1 = (1 + r^2)^{-1/2} r \sin t \, dr + (1 + r^2)^{1/2} \cos t \, dt$$

$$dT_2 = (1 + r^2)^{-1/2} r \cos t \, dr - (1 + r^2)^{1/2} \sin t \, dt$$

$$dX_1 = dr \sin\theta \cos\phi + r \cos\theta \cos\phi \, d\theta - r \sin\theta \sin\phi \, d\phi$$

$$dX_2 = dr \sin\theta \sin\phi + r\cos\theta \sin\phi d\theta + r\sin\theta \cos\phi d\phi$$

$$dX_3 = dr \cos\theta - r\sin\theta d\theta.$$

We can now compute the metric on the surface (6.6):

$$
\begin{aligned}
d\tau^2 &= dT_1^2 + dT_2^2 - dX_1^2 - dX_2^2 - dX_3^2 \\
&= [(1+r^2)^{-1/2} r\sin t\, dr + (1+r^2)^{1/2} \cos t\, dt]^2 \\
&\quad + [(1+r^2)^{-1/2} r\cos t\, dr - (1+r^2)^{1/2} \sin t\, dt]^2 \\
&\quad - [dr \sin\theta \cos\phi + r\cos\theta \cos aphid\theta - r\sin\theta \sin\phi d\phi]^2 \\
&\quad - [dr \sin\theta \sin\phi + r\cos\theta \sin\phi d\theta + r\sin\theta \cos\phi d\phi]^2 \\
&\quad - [dr \cos\theta - r\sin\theta d\theta]^2
\end{aligned}
$$

from which

$$
\begin{aligned}
d\tau^2 &= \frac{r^2}{1+r^2} dr^2 + (1+r^2)dt^2 - dr^2 \\
&\quad - r^2\cos^2\theta d\theta^2 - r^2\sin^2\theta d\phi^2 + r^2\sin^2\theta d\theta^2 \\
&= (1+r^2)dt^2 - \frac{dr^2}{1+r^2} - r^2 d\theta^2 - r^2\sin^2\theta d\phi^2.
\end{aligned}
$$

Reinstating l gives (6.2).

Problem 6.2. If $a = 0$ then $r_+ = l\sqrt{m}$ and $r_- = 0$ from (6.11). Then substitution in (6.10) and a little rearrangement gives the result.

Problem 6.3. Multiplying through by r we get

$$\frac{d}{dr}\left(r\frac{d\varphi}{dr}\right) = GM\delta(r) + \left(\frac{2}{l}\right)^2 r.$$

Integrating, we get

$$r\frac{d\varphi}{dr} = GM + \left(\frac{2}{l}\right)^2 \frac{r^2}{2} + c.$$

Take $c = -GM$. Then

$$\frac{d\varphi}{dr} = \left(\frac{2}{l}\right)^2 \frac{r}{2}.$$

Integrating again gives $\varphi = r^2/l^2 + c$ so we choose $c = -m$ to obtain the required result.

Problem 6.4. In 4 spacetime dimensions we have $\mu = GM$, $\Omega_2 = 4\pi$ and $d\Omega_0 = 0$ from which the result follows.

Problem 6.5. (a) The entropy is $S_{\rm bh} = A_h/4G_5$ where A_h is the horizon area. For five dimensions we have $\Omega_3 = 2\pi^2/\Gamma(2) = 2\pi^2$ so the area $A = 2\pi^2 r_5^3$. Thus $S_{\rm bh} = \pi^2 r_5^3/(2G_5)$.

(b) From the metric equation (6.12) the horizon ($f = 0$) is at $r_5^2 = 16\pi G_5 M/(3\Omega_3) = 8G_5 M/3\pi$.

(c) In the four-dimensional spacetime orthogonal to the string the metric is Schwarzschild. So the horizon radius is $r_h = 2GM_s$ (where as always $G \equiv G_4$) and the entropy is $S_{\rm bs} = A/4G = \pi r_h^2/G = 4\pi GM^2$.

If now $M = M_s = r_h/2G$ and $G_5 = LG$ then

$$\frac{S_{\rm bh}}{S_{\rm bs}} = \frac{\pi^2}{2GL}\left(\frac{8GLM}{3\pi}\right)^{3/2} \times \frac{1}{4\pi M^2 G}$$

$$= \left(\frac{8L}{27\pi M}\right)^{1/2}.$$

Thus for large enough L the black hole will have a higher entropy so the string is likely to be unstable. (Numerical computations confirm this.)

Problem 6.6. We have

$$\kappa = \frac{1}{2}\frac{dg_{00}}{dr}\bigg|_{r=r_+} = -\frac{1}{2}\left[\frac{2m}{r_+^2} - \frac{2}{r}\frac{dm}{dr}\right]_{r=r_+}$$

where

$$\frac{dm}{dr} = \frac{1}{b^2}\left(\sqrt{r^4 + b^2 Q^2} - r^2\right).$$

Now, $g_{00} = 0$ when $r_+ = 2m(r_+)$, so this gives the horizon radius (when a solution exists). Then

$$\kappa = \frac{1}{2r_+} - \frac{1}{b^2 r_+}\left(\sqrt{r_+^4 + b^2 Q^2} - r_+^2\right).$$

Varying r_+ (and hence the horizon area) and the charge Q, we get

$$dM = dm - \frac{1}{b^2}\left(\sqrt{r_+^4 + b^2 Q^2} - r_+^2\right)dr_+ + \frac{1}{b^2}\int_{r_+}^{\infty}\frac{Qb^2 dQ}{\sqrt{x^4 + b^2 Q^2}}dx$$

$$= r_+ dr_+\left[\frac{1}{2r_+} - \frac{1}{r_+}\left(\sqrt{r_+^4 + b^2 Q^2} - r_+^2\right)\right] + \Phi_h dQ$$

$$= \frac{\kappa}{8\pi}dA + \Phi_h dQ$$

where $dA = d(4\pi r_+^2) = 8\pi r_+ dr_+$.

Problem 6.7. For $z = 1$ the Lifshitz black hole metric reduces to

$$d\tau^2 = \frac{r^2}{l^2}\left(1 - \frac{ml^2}{r^2}\right)dt^2 - \frac{l^2}{r^2}\left(1 - \frac{ml^2}{r^2}\right)^{-1}dr^2 - r^2 d\phi^2,$$

which gives the BTZ metric in the form (6.10) with a little rearrangement. The black hole horizon is $r_+ = l\sqrt{m}$. For general z

$$\sqrt{-g^{00}g^{11}}\,\bigg|_{r=r_+} = \left(\frac{l^{2z-2}}{r_+^{2z-2}}\right)^{1/2} = \frac{1}{m}$$

and

$$\frac{dg_{00}}{dr}\bigg|_{r=r_+} = \frac{r_+^{2z}}{l^{2z}}\frac{2ml^2}{r_+^3} = \frac{2m^{z-1/2}}{l}.$$

Thus $T = \kappa/2\pi = m^{z-3/2}/2\pi l$. For $z = 3$ this is $m^{3/2}/2\pi l$. The entropy is $S = A/G_3 = 2\pi r_+/G_3$. So

$$dM = \frac{r_+^3}{G_3 l^4}dr_+ = \frac{r_+^3}{2\pi l^4}dS = \frac{m^{3/2}}{2\pi l}dS = T dS$$

as required.

Problem 6.8. We have

$$d\tau^2 = \frac{r^2}{R^2}\left[\left(1 - \left(\frac{r_0}{r}\right)^4\right)dt^2 - dw^2 - dx^2 - dy^2\right]$$

$$-\frac{R^2}{r^2}\left[1 - \left(\frac{r_0}{r}\right)^4\right]dr^2 - R^2 d\Omega_5^2$$

$$\sim \frac{r^2}{R^2}(dt^2 - dw^2 - dx^2 - dy^2) - \frac{R^2}{r^2}dr^2 - R^2 d\Omega_5^2$$

as $r \to \infty$. Comparing this to equation (6.8), we see that, with a change of notation ($\tilde{t} \to t$, $\rho \to r$, $x_2 \to x$, $x_3 \to y$, $l \to R$), and the addition of the w coordinate, the coordinates (t, r, w, x, y) describe an AdS metric in five dimensions. The remaining coordinates (contained in $d\Omega_5^2$) represent a sphere in five dimensions. The radius of curvature of the sphere and of the anti-de Sitter spacetime are both R.

Problem 6.9. The metric is

$$d\tau^2 = \left[1 - \left(\tfrac{r_1}{r}\right)^{d-3}\right]\left[1 - \left(\tfrac{r_2}{r}\right)^{d-3}\right] dt^2 \tag{17}$$

$$- \left[1 - \left(\tfrac{r_1}{r}\right)^{d-3}\right]^{-1}\left[1 - \left(\tfrac{r_2}{r}\right)^{d-3}\right]^{-1} dr^2. \tag{18}$$

For the near-horizon metric put

$$r = r_1 + \alpha\epsilon^2$$

where α is a constant to be chosen and ϵ is small. Then, to lowest order in ϵ

$$d\tau^2 = \frac{(d-3)\alpha\epsilon^2}{r_1}\left[1 - \left(\frac{r_2}{r_1}\right)^{d-3}\right] dt^2 - \frac{\alpha r_1}{4(d-3)}d\epsilon^2.$$

Taking

$$\alpha = \frac{(d-3)}{4r_1}\left[1 - \left(\frac{r_1}{r_2}\right)^{d-3}\right] \quad \text{and} \quad \theta = 2\alpha$$

gives

$$ds^2 = \epsilon^2 d\theta^2 + d\epsilon^2.$$

If θ now has period 2π, t has period $2\pi/2\alpha$ and the temperature of the black hole is therefore α/π or

$$T = \frac{(d-3)}{4\pi r_1}\left[1 - \left(\frac{r_1}{r_2}\right)^{d-3}\right].$$

If $r_1 = r_2$ we get $T = 0$ independent of d. For $d = 4$ we have

$$T = \frac{r_1 - r_2}{4\pi r_1^2}$$

where $r_1 = m + (m^2 - q^2)^{1/2}$ and $r_1 - r_2 = 2(m^2 - q^2)^{1/2}$. Multiplying the numerator and denominator of T by r_1 and substituting for r_1 and $r_1 - r_2$ in the numerator gives the required result.

Problem 6.10. Comparing the metric (6.27) with that of the 3-sphere we can indentify $R^2 = f^{1/3}r^2$. Thus

$$R^2 = f^{1/3}r^2 = \frac{1}{r^2}(r^2 + r_1^2)^{1/3}(r^2 + r_2^2)^{1/3}(r^2 + r_3^2)^{1/3} \to (r_1 r_2 r_3)^{2/3}$$

as $r \to 0$ at the horizon. It follows that the area of the horizon is $2\pi^2 R^3 = 2\pi^2 r_1 r_2 r_3$.

Problem 7.1. (a) From Section 3.13.2 the orbit must be prograde since r_{ms} for a retrograde orbit is $9m$. We can solve equation (3.59) for a (see problem 3.18) as a function of r_{ms}. Putting $r_{ms} = 1.2m$ gives $a = .9987m$. Given the limited accuracy of the data this is at the limit implied by Thorne (Section 3.16.1).

(b) The energy in the orbit as seen from infinity, E, is given by (equation. (3.58))

$$E = \left(1 - \frac{2m}{3r_{ms}}\right)^{1/2} = 0.67.$$

The binding energy is $1 - 0.67 = 0.33$ per unit particle rest mass.

Problem 7.2. As both stars are in circular orbits, from Newton's second law,

$$\frac{\nu_s^2}{r_s} = \frac{GM_d}{r^2}. \tag{19}$$

From the definition of the centre of mass of the system, $M_s r_s = M_d r_d$, therefore the distance r between the two stars is

$$r = r_s \left(1 + \frac{M_s}{M_d}\right). \tag{20}$$

The orbital period of the system is

$$P = 2\pi r_s / \nu_s. \tag{21}$$

Combining (19) and (20) gives $r_s = \frac{GM_d}{\nu_s^2(1+M_s/M_d)^2}$ and substituting for r_s from (21) gives $\frac{P\nu_s^3}{2\pi G} = \frac{M_d}{(1+(M_s/M_d)^2)}$. But ν_s is the velocity in the plane of the orbit; this is related to the line of sight velocity by $V_s = \nu_s \sin \phi$, so finally

$$\frac{PV_s^3}{2\pi G} = \frac{M_d \sin^3 \phi}{(1 + M_s/M_d)^2}.$$

References

Chapter 1

Hartle J 2003 *Gravity, An Introduction to Einstein's General Relativity* (Addison Wesley).

Kenyon I R 1990 *General Relativity* (Oxford University Press).

Will C 1993 *Theory and Experiment in Gravitational Physics* (Cambridge University Press, Cambridge UK).

Chapter 2

Doeleman S S *et al.* 2008 *Nature* **455** 78.

Frolov V P and Novikov I D 1998 *Black Hole Physics* (Kluwer, Dordrecht).

Hamilton A J S and Lisle J P 2008 *Am J Phys* **76** 519–532.

Oppenheimer J R and Snyder H 1939 *Phys Rev* **56** 45.

Taylor J H 1994 *Rev Mod Phys* **66** 711.

Taylor E F and Wheeler J A 2000 *Exploring Black Holes* (Addison Wesley Longman).

Thorne K S, Price R H and Macdonald D A 1986 *Black Holes: The Membrane Paradigm* (Yale University Press, New Haven).

Chapter 3

Bardeen J M 1970 *Nature* **226** 64.

Bardeen J M *et al.* 1972 *Astrophys J* **178** 347.

Ciufolini I *et al.* 2007 *Int J Mod Phys D* **16** 2271–2285.

Frolov V P and Novikov I D 1998 *Black Hole Physics* (Kluwer, Dordrecht/ London).

Lynden-Bell D 1978 *Physica Scripta* **17** 185.

Thorne K S 1974 *Astrophys J* **191** 507.

Townsend P *Black Holes*, www.damtp.cam.ac.uk/lecturenotes/BH.ps (unpublished lecture notes).

Chapter 4

Bekenstein J D 1975 *Phys Rev* **D12** 3077.

Bekenstein J D 1980 *Black Hole Thermodynamics Physics Today* **24**.

Birrell N D and Davies P C W 1982 *Quantum Fields in Curved Spacetime* (Cambridge University Press).

Carlip S 2009 in E Papantonopoulos (ed) *Physics of Black Holes: A Guided Tour* (Lecture Notes in Physics 769, Springer).

Chandrasekhar S 1983 *The Mathematical Theory of Black Holes* (Oxford University Press).

Custodio and Horvath 2003 *Am J Phys* **71** 1237.

Frolov V P and Novikov I D 1998 *Black Hole Physics* (Kluwer, Dordrecht/ London).

Futterman J A H, Handler F A and Matzner R A 1988 *Scattering from Black Holes* (Cambridge University Press, Cambridge UK).

Hawking S W 1974 *Nature* **248** 30.

Hawking S W 1976 *Phys Rev* **D13** 191.

Keifer C 1999 in *Classical and Quantum Black Holes* eds P Fré, V Gorini, G Magli and U Moschella (Institute of Physics, Bristol).

Takagi S 1986 *Vacuum Noise and Stress Induced by Uniform Acceleration*, Prog Theor Phys Supp **88** 1.

Townsend P K *Black Holes*, www.damtp.cam.ac.uk/lecturenotes/BH.ps (unpublished lecture notes).

Unruh 1976 *Phys Rev* **D14** 870.

Chapter 5

Bordag M, Mohideen U and Mostepanenko V M 2001 *Physics Reports* **353** 1–205.

Jensen B, McLaughlin J and Ottewill A 1988 *Class Quantum Grav* **5** L187-L189.

Morris M S and Thorne K S 1988 *Am J Phys* **56** 395.

Morris M S, Thorne K S and Yurtsever U 1988 *Phys Rev Lett* **61** 13.

Thorne K S 1994 *Black Holes and Time Warps: Einstein's Outrageous Legacy* (Picador).

Visser M 1995 *Lorentzian Wormholes* (AIP).

Chapter 6

Bañados M, Henneaux M, Teitelboim C and Zanelli J 1993 *Phys Rev* **D48** 1506–1525.

Becker K, Becker, M and Schwarz J H 2007 *String Theory and M-Theory: A Modern Introduction* (Cambridge).

Carlip S 2005 *Conformal Field Theories, (2 + 1)-Dimensional Gravity and the BTZ Black Hole Class Quant Grav* **22** R85-R124 and arXiv:gr-qc/0503022.

Carlip S 2009 in E Papantonopoulos (ed) *Physics of Black Holes; A Guided Tour* (Lecture Notes in Physics 769, Springer).

Gregory R 2009 in E Papantonopoulos (ed) *Physics of Black Holes; A Guided Tour* (Lecture Notes in Physics 769, Springer).

Malcedena J M 1998 *Adv Theor Math Phys* **2** 231–252 see also Petersen J L 1999 *Introduction to the Maldacena Conjecture on AdS/CFT Int J Mod Phys* **A14** 3597–3672.

Mann R B 2008 *Topological Black Holes-Outside Looking In* arXiv:gr-qc/9709039.

Meyers R C and Perry M J 1986 *Ann Phys* **172** 304–2347.
Myung Y S 2012 *Phase transitions for the Lifshitz black holes* arXiv:1203. 1367[hep-th].
Rasheed D A 2008 *Non-linear Electrodynamics* arXiv:hep-th/9702087.
Zerilli F *J Phys Rev DZ* 1970 2141–2160.

Chapter 7

Bardeen J M and Pettersen J A 1975 *Astrophys J* **105** L65.
Barkana L H and Loeb A 2003 *Nature* **421** 341.
Begelman M and Rees M 1998 *Gravity's Fatal Attraction* (Scientific American).
Burgay M *et al.* 2003 *Nature* **426** 531.
Centrella J *et al.* 2010 *Rev Mod Phys* **82** 3069.
Cottam J *et al.* 2002 *Nature* **420** 51.
Doeleman S S *et al.* 2008 *Nature* **455** 78–80.
Gebhardt K, Rich R M and Ho L 2002 *Astrophys J* **L41** 578.
Genzel R *et al.* 2003 *Nature* **425** 934.
Hawking S W and Ellis G F R 1973 *The Scale Structure of Spacetime* (Cambridge University Press).
Hjorth J *et al.* 2003 *Nature* **423** 847.
MacGibbon J and Carr B 1991 *Astrophys J* **371** 447.
McLintock J E *et al.* 2003 *Astrophys J* **593** 435.
Melia F and Falcke H 2001 *Ann Rev Astron Astrophys* **39** 342
Meyer L *et al.* 2012 *Science* **338** (6103): 84–87.
Orosz J A *et al.* 2007 *Nature* **449** 872.
Orosz J A *et al.* 2011 *Astrophys J* **742** 84.
Price R H 1972 *Phys Rev* **D5** 2410.
Taylor J H 1994 *Rev Mod Phys* **66** 711.
Van Der Marel R P *et al.* 2002 *Astron J* **124** 3255.
Zhang S N and Cui W 1998 *Astrophys J* **482** L155.

Bibliography

The following is a selection of books on black hole physics at various levels.

Thorne K S 1994 *Black holes and time warps* (Papermac): gives a non-mathematical historical survey and includes the problems of time travel.

Luminet J-P 1987 *Black holes* (Cambridge University Press): a semi-popular exposition.

Susskind L and Lindsay J 2005 *An introduction to Black Holes, Information and the String Theory Revolution: The Holographic Universe* (World Scientific): addresses key issues using only undergraduate level mathematics.

Taylor E F and Wheeler J A 2000 *Exploring black holes* (Addison Wesley Longman): an introduction for those with a basic knowledge of general relativity but without using tensor calculus.

Lightman A P *et al.* 1974 *Problem book in relativity and gravitation* (Princeton): problems for advanced students (with outline answers).

Fre P (ed) *et al.* 1999 *Classical and quantum black holes* (Institute of Physics): monograph on current developments in black holes and superstrings with an introductory first chapter.

Chandrasekhar S 1983 *The Mathematical theory of black holes* (Oxford): comprehensive mathematical treatment of classical black hole spacetimes.

Frolov V P and Zelnikov A 2011 *Introduction to Black Hole Physics* (Oxford): detailed survey of black holes including the mathematics of general relativity and quantum aspects. This covers similar material to that of Frolov V P and Novikov I D (1998 *Black Hole Physics* (Kluwer, Dordrecht)) which we refer to in the text and is now out of print.

Calmet X, Carr B and Winstanley E (2014) Quantum Black Holes (Springer): a brief survey covering much of our material at a somewhat more advanced level.

Frank J, King A R and Raine D J 2002 *Accretion power in astrophysics* (Cambridge University Press): astrophysics of gas flows round compact objects.

Papantonopoulos E (ed) 2009 *Physics of Black (Holes; A Guided Tour* (Lecture Notes in Physics 769, Springer): a collection of lectures covering contemporary issues in black hole thermodynamics, AdS black holes, higher dimensions, alternative theories, colliding black holes and gravitational radiation mosty at graduate level.

Horowitz G T (ed) 2012 *Black Holes in Higher Dimensions* (Cambridge): an advanced text on recent issues and results.

Index

Printed in the United States
By Bookmasters